Energy

Modern Life, Climate Change, and Oil Production

Robert L. Hirsch, Ph.D.

Apogee Prime

Energy - Modern Life, Climate Change and Oil Production

ISBN 1-978-1926837-43-7
First Edition

Parts of this book appeared in the volume "The Impending World Energy Mess" by Hirsch, Wendling and Bezdek, published in 2010.

Apogee Prime is an imprint of CG Publishing Inc. Burlington, Ontario, Canada, L7R 2B5 www.apogeeprime.com

Is this book for you?

- **Yes,** if you're interested in important energy realities, many of which will impact you.
- **Yes,** if you're interested in learning about the essential products derived from oil and gas (think beyond gasoline to pharmaceuticals).
- **Yes,** if you want to understand the dangers associated with many on-going political decisions, independent of political party.
- **Yes,** if you want to understand the benefits and shortcomings of various energy technologies … Hint: None are perfect.
- **Yes,** if you're interested in solutions to current and future problems.
- **No**, if you believe the current approach to energy is satisfactory.
- **No**, if you believe that the U.S. must decarbonize, independent of China and India.
- **No,** if you want to ignore the experience of others in decarbonization.
- **No,** if you're not interested in lurking world oil supply issues.

Contents

Foreword		7
Chapter I	Introduction	9
Chapter II	Energy in Your Workday	11
Chapter III	Products From Oil/natural Gas, A Boone to Humankind	15
Chapter IV	A Brief History of Energy	19
Chapter V	Climate Change Enlightenment & Stupidity	23
Chapter VI	Some Perspectives on Oil	27
Chapter VII	Finding and Producing Oil	43
Chapter VIII	Oil Field Reserves	53
Chapter IX	How Long Adequate Oil?	59
Chapter X	Peak Oil Mitigation	63
Chapter XI	Recommendations	69
Chapter XII	Conclusions	71
Appendix I.	Thoughts from Dr. James R. Schlesinger	73
Appendix II.	Thinking About Our Energy System	75
Appendix III.	The World Oil Enterprise	77
Appendix IV.	More on Oil Production	81
Appendix V.	Oil Production Dynamics	85
Appendix VI.	How Might an Oil Debacle Unfold?	89
Appendix VII.	Saving Our Way Out of Impending World Oil Shortages	91
Appendix VIII.	Other Transportation Fuels	93
Appendix IX.	Administrative Mitigation	103
Appendix X.	The Best That Physical Mitigation Can Provide	117
Appendix XI.	What Forecasters Have Forecast	131
Appendix XII.	Electric Power Options	137
Appendix XIII.	The Russia-Ukraine War: A Dry, Dry Run-on Peak Oil?	143
Acknowledgements		148
Index		149
About the Author		150

Foreword

The subject of energy is difficult for many to understand, so it can't be addressed in a few paragraphs. In the following you'll be carried through your typical day, identifying important energy and energy byproducts on an almost hour-by-hour basis. Next comes the major issue of global climate change and what I believe is the best way to attack the problem. Finally, the issue of the impending world oil shortage is addressed with its unknown timing and what it will take to best mitigate that world-shaking problem. Realities and meaningful options are discussed on a factual basis, letting the politics fall where they may.

When it comes to the subject of world oil production, some of the mechanics of finding and developing an oil field are described and extrapolated to the world scale, keeping it as simple as I know how. You'll learn that there are two kinds of world maximum world oil production lurking. Both are almost certain to be catastrophic for you. Forecasting future maximum world oil production is almost impossible, but it wouldn't be a surprise if an underinvestment-related world oil production shortage occurs relatively soon.

This book is a substantial update of our 2010 book entitled "The Impending World Energy Mess," which I coauthored with Roger Bezdek and Bob Wendling. In this edition I do my best to bring energy down to the non-expert level. Significant parts of our earlier book are still relevant, so they've been updated and provided as appendices. You decide which detailed appendix may interest you.

Hopefully, you'll benefit from these pages, make better energy-related decisions, and make your voice heard, where appropriate. It's your country and your life. Both depend vitally on energy.

Chapter I. Introduction

Energy and its by-products dominate your life in more ways than you might imagine; energy is much more than gasoline for your car or electric power for your home. Energy is not a simple or easy subject in spite of what many would have you believe; politics and ideology too often cloud and distort related public discourse. Furthermore, technical people often talk "technical," which can leave non-technical people confused.

Talking about energy as an overall category can be confusing and sometimes outright misleading. Energy discussions should be broken down into **LIQUID ENERGY, SOLID ENERGY, GASEOUS ENERGY, and SOLAR ENERGY.**

It was recently suggested that the United States was "energy independent." That was correct only in that the NET energy into the U.S. was near zero on an **energy content** point of view, meaning export energy was about equal to imported energy. But the U.S. situation was really out of balance, because **the U.S. was and is importing roughly a third of its oil needs**. Put another way, if all oil imports were suddenly stopped in this so-called "energy independent" situation, the U.S. would be tragically short of oil.

In the following pages, energy and some of its wonders will be discussed. To give you some background, a short history of U.S. energy evolution will be provided. In the past, energy choices were determined primarily by evolution and economics, not politicians.

As this book goes to press, **widespread concerns about global climate change have led governments to intervene in the energy marketplace**. Some want to stop world oil production to "save the planet" from global warming. The problem is that energy machinery (cars, truck, airplanes, trains, factories, etc.) represent an enormous investment that cannot be quickly replaced. In addition, **some amount of oil will remain essential to our very existence, because of what it provides in the way of high value products and pharmaceuticals.**

Concerns about global climate change are pervasive and cannot be ignored. Nevertheless, actions by some governments have been foolhardy (some might say "stupid") because they sacrifice industrial base (read: jobs) and burden individual families with higher energy bills. There are more intelligent ways to proceed. These issues are impacting now, and you need to recognize them.

Concerns with climate change have impacted world oil production for years, resulting in significant reductions in investments in oil exploration and production. Since investment is a critical precursor to future oil production, those well-intentioned interventions translate to lower future oil production, so **it's conceivable that world oil production will not meet demand in the near-term, resulting in dramatic upsets to our daily lives and economies**. "Near-term" is defined here as the time needed to start essential mitigation, which you will see is measured in decades.

The investment-related oil supply disruption will be followed at some future date with what we call **geologically-dictated oil supply shortages, which are certain to lead to**

incredible human and economic suffering. How these disasters might unfold are described, along with how governments might attempt to soften the blows, short-term and long-term.

Let me reiterate … **The problem will be oil, not the broad category of "energy**;" wind mills provide electric power which cannot power your gasoline engine automobile.

Finally, an extensive number of appendices are provided in which there's greater detail on aspects of the anticipated problems that exist now and lie ahead. As an example, if the sudden U.S. oil shortages that occurred in 1973 and 1979 provide real-world experience on what governments might do in the future. Much of what happened in years past will be of great value in addressing the oil shortage challenges that lie ahead.

Chapter II. Energy in Your Workday

Introduction

Energy and energy byproducts are everywhere in our lives, but people often don't recognize them. In the following, energy and related by-products are described in your energy day. We break energy forms down to **electric** or **electricity**, which comes from a number of sources, and oil/natural gas (**oil/nat. gas**), which yield a wide range of products. We'll expand on some of this later, but first, let's see how these energy products impact some of our readers.

Morning

You're awakened by your **electric** alarm clock made in part from **oil/nat. gas**. You turn on your **electric** light and head for a hot shower, powered by **electricity or natural gas**, using soap derived from **oil/nat. gas**. You brush your teeth with a tooth brush and toothpaste, ingredients of which came from **oil/nat. gas**.

Breakfast might be oatmeal and milk, both produced and brought to you via **electricity** and **oil/nat. gas** (cows are milked electrically). Your orange juice was harvested and delivered via **oil/nat. gas**. You get dressed in clothes partially made and delivered by **oil/nat. gas** from stores or warehouses lit by **electricity**. Next, you might get in your car, which is partially constructed from **oil/nat. gas** and powered in most cases by **oil/nat. gas**. You drive on roads made of asphalt or concrete produced from or using **oil/nat. gas**, controlled by traffic lights, powered by **electricity**. Oops! We forgot to check your cell phone (**electric**), partially made from **oil/nat. gas**. And all the electric wiring in your abode, what you can see and what's in the walls, is covered with insulation (**oil/nat. gas).**

Next, we briefly consider a number of jobs and the energy forms involved: office worker, truck driver, doctor, stay-at-home parent, and office worker.

Local Truck Driver

Let's assume you drive a delivery truck locally. You get to your truck, which is partially constructed of **oil/nat. gas** and fueled by **oil/nat. gas** (not many electrics at this writing). You check your cell phone (**electric, oil/nat. gas**) to see how your supervisor wants you to spend your day, what to pick up where and what to deliver somewhere else. Your loads are likely partially produced and or delivered to your pickup locations by **oil/nat. gas**. You deliver to locations constructed in part using **oil/nat. gas** and air conditioned by **electricity**. Offloading is done by battery powered (**electric**) fork lifts.

You're off for your next load, stopping for fuel first (**oil/nat. gas**). Along the way you stop for a cup of coffee, imported by **oil/nat. gas**, and served from a shop heated or cooled by **oil/nat. gas** or **electricity**. At noon you might eat from a bagged lunch or stop at a fast-food place, whose food materials were delivered by refrigerated trucks (**oil/nat. gas**), stored in large refrigerators (cooled by **electricity**), cooked on **electric or oil/nat. gas** appliances, and provided to you. You eat in a room heated or cooled (**electricity** or **oil/nat. gas**) at tables partly constructed of **oil/nat. gas** and maybe watch television (**electricity**). You dump your refuse in a trash bin made from **oil/nat. gas**. It's eventually removed by truck (**oil/nat. gas**) to a dump, where part of it is incinerated with **nat. gas.** Let's assume your afternoon is similar to your morning. We'll deal with your evening after we consider some other vocations.

Doctor

You may first drive to the hospital for your rounds (**oil/nat. gas**). Your computer or smartphone (**electricity** and **oil/nat. gas**) gives you patient information, which you confirm and/or alter. Some of your patients are connected to medical equipment run by **electricity** and made in part from **oil/nat. gas**. Maybe you prescribe medications (**oil/nat. gas**).

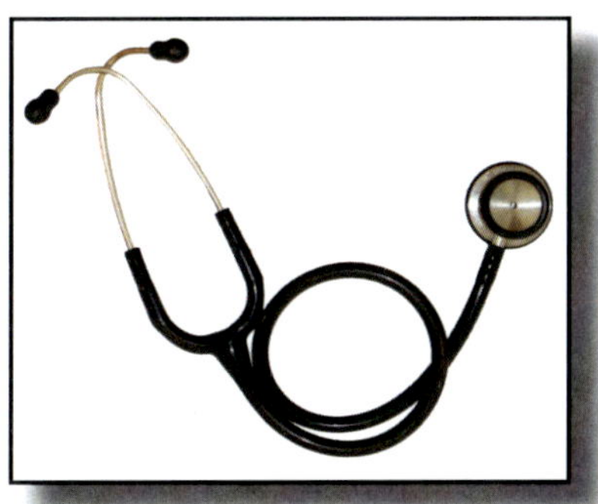

When your rounds are complete, let's say you're off to your office in your car (**oil/nat. gas**). Your office is powered by **electricity** and built in part with **oil/nat. gas**-derived materials. Your day is mapped out on your computer (**electricity** and **oil/nat. gas**). You diagnose disease during the day with various equipment (**electricity** and **oil/nat. gas**) and prescribe various medications (**oil/nat. gas**). On occasion you're called on your cell phone (**electricity** and **oil/nat. gas**) because other doctors need your advice or patients need quick assistance.

Stay-At-Home Parent

After breakfast you put your dishes into the dishwasher (**electricity** and **oil/nat. gas**), and play with your young kids with their various toys (**electricity** and **oil/nat. gas**). Maybe they want to play with friends elsewhere, so you put them in the car (**oil/nat. gas**) and drive them. Maybe you stay to help with the bedlam.

After, it's a drive back home (**oil/nat. gas**), where you feed them lunch in part from the refrigerator (**electricity** and **oil/nat. gas**). Then, maybe it's their nap time, which gives you some time to read a book (**electricity** and **oil/nat. gas**) in your air-conditioned home (**electricity**), maybe with a cup of coffee (**electricity** and **oil/nat. gas)** Maybe a bath to relax (**electricity** and **oil/nat. gas**) before making dinner (**electricity** and **oil/nat. gas**).

Office Worker

No matter what your level, you probably work in an air-conditioned office building (**electricity** and **oil/nat. gas**). You get to your desk with your computer (**electricity** and **oil/nat. gas**), check your must-do's, should-do's, and ought-to-do's. Maybe you grab a cup of coffee (**electricity** and **oil/nat. gas**). At break, maybe it's coffee again (**electricity** and **oil/nat. gas**), while talking to colleagues.

At noon you may eat lunch in your cafeteria (**electricity** and **oil/nat. gas**). Maybe you run an errand in your car (**electricity** and **oil/nat. gas**). Late in the day it's off to home.

End of the Day

You drive home in your car, made partially from **oil/nat. gas**, and powered by **oil/nat. gas**. Before dinner you might have a beer from your refrigerator (**electric**), that came to you via train, truck (**oil/nat. gas, electricity**), and your local grocery store. Maybe you chat with your spouse and kids. The fixings for dinner came from the country, where they were grown in fertilized fields (**oil/nat. gas**) and harvested (**oil/nat. gas**), processed (**oil/nat. gas and electricity**), brought to your local grocery (**oil/nat. gas and electricity**) and brought to your home, where most were kept in your refrigerator (**electricity**). Cooking might be on an electric cooking area or microwave (**electricity**) or a gas stove (**oil/nat. gas**).

After dinner the dishes might go into a dishwasher made in part from **oil/nat. gas** and powered by **electricity**. Maybe you watch television (**electricity**) or read via a lamp (**electricity**) before going to bed. You wash up with warm water (**oil/nat. gas** or **electricity**), brush your teeth (**oil/nat. gas**), set your alarm clock, made in part from **oil/nat. gas** and powered by **electricity**, settle onto your mattress (**oil/nat. gas**) turn off the lights (**electricity**) and go to sleep.

Chapter III. Products From Oil/natural Gas, A Boone to Humankind

Figure III-1 provides a rough picture of the major products made from a barrel of oil. Most is for fuels (80%), some for road paving, roof shingles, and a myriad of related products (asphalt at 3%), and the rest can be lumped into miscellaneous. These numbers are reasonable; just be aware that oils from different locations can be somewhat different in composition. Those differences lead to different prices for different oils, and different oil refineries are optimized to take particular oils. Skipping over related details doesn't really impact your overall understanding. **Just keep in mind that there is a great deal of detail that's important in the business but less important in the overall picture.**

Oil/petroleum as it comes from the ground

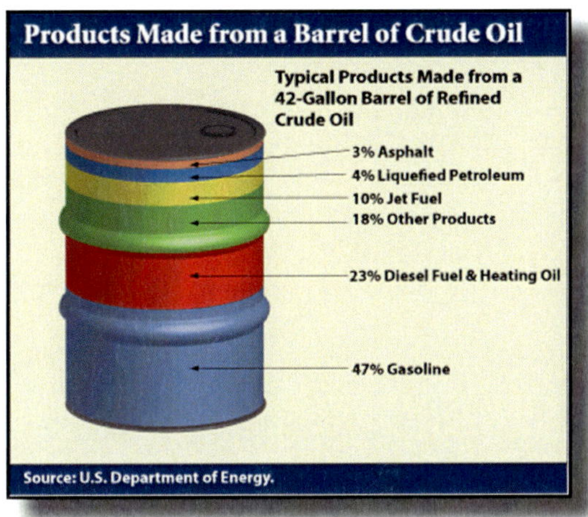

Figure III-1. What a typical barrel of oil provides.

We all recognize that gasoline is essential to fuel the vast majority of automobiles and light trucks in the U.S. Heavy trucks, most buses, and industrial equipment run on diesel engines, which require diesel fuel, which has a different composition than gasoline. And there are airplanes running on jet fuel. Then there's motor oils, which lubricate engines, and heating oil that's used in more isolated, colder areas of the country for home heating. If you haven't done so recently or at all, spend some time looking around you and listing all of the machines in the transportation sector that run on liquid fuels. It's really quite amazing.

Sometimes natural gas is involved in refineries, sometimes not. Neglecting its refinery use doesn't lose much in our basic understanding. In general, natural gas is used directly in various applications, such as heating of buildings, heating of water, industrial applications, etc.

Under "Other Products" in the Figure III-1 is an array of high value items made from oil/natural gas that are everywhere in our everyday lives. Figure III-2 shows a partial listing.

Figure III-2. Some of the products we get from oil and natural gas.

Probably most important is the vast array of pharmaceuticals, which modern society can't live without. Some of the other important products derived at least in part from oil/natural gas are the following:

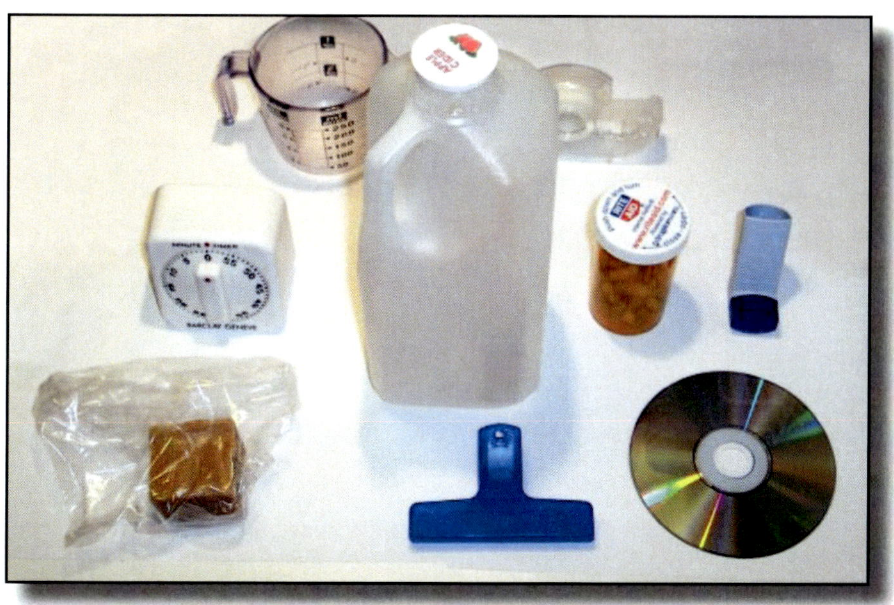

Plastic products

- Plastics
- Ink
- Paint
- Shoe Polish
- Roof shingles
- Cosmetics
- Candles
- Vaseline
- Bug Killer
- Ammonia
- Tires
- Paper cups
- Wax paper
- Upholstery
- Vitamin capsules
- Footballs
- Deodorant
- Trash bags

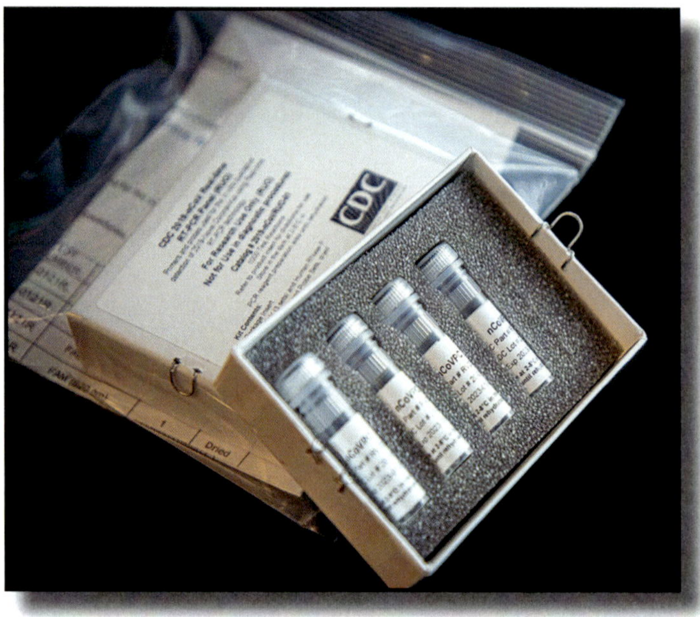

Pharmaceutical covid test kit

And the list goes on. Hopefully, this helps you realize the extent of oil products in our lives. If you want more, Google oil products; you'll be amazed.

Plastics? You may not love plastics because they're part of the ugly pollution on the land and in lakes, rivers and oceans. There's no question that improperly discarded plastics can be a pollution problem. **But plastics' pollution is a human problem, yet to be satisfactorily addressed.** If you want to rid the world of plastics, first define what you're going to replace them with, at what cost, and with what waste management problems. There are few viable alternatives.

The bottom line is that very many products from oil/natural gas are essential to modern living. Maybe they could be synthesized some other way, but doing so is sure to be more expensive. It's safe to assume that **as long as there's some oil/natural gas in the world, these high-value products will be given preference for production**. At some distant future date, we may be living in a world where oil and natural gas are no longer adequately available, but many of these high-value products will continue to receive priority for what's left.

Chapter IV. A Brief History of Energy

The history of human development is a history of human ability to harness various forms of energy. **Early humans satisfied their needs with their own physical capabilities. Then, along came fire roughly 1.5 million years ago.** Fire was a source of heat and light, protection from predators, and a means of cooking. Later, fire was used to make hunting tools. It's really difficult sitting in our homes, schools, or offices today to imagine basic humans conquering fire to advance to a very primitive style of life.

Agriculture is said to have started about 12,000 years ago. The National Geographic Society tells us, "Traditional hunter-gatherer lifestyles, followed by humans since their evolution, were swept aside in favor of permanent settlements and a reliable food supply. Out of agriculture, cities and civilizations grew, and because crops and animals could now be farmed to meet demand, the global population rocketed — from some five million people 10,000 years ago, to more than seven billion today."

As humans evolved, they found that they could use animals to magnify their efforts and enhance their lives. Various animals were domesticated. This milestone occurred in Asia around 8,000 BCE. Sheep and goats were tamed and raised for their meat, milk and hides. Oxen, horses, and donkeys were used for carrying and pulling heavy loads. This process evolved over a long period of time.

Along the way fossil fuels began to appear. Maybe 3,000 years ago, the Chinese were believed to have started to use coal for heat and light. In this hemisphere during the 1300s, the Hopi Indians used coal that was readily available.

Around 1760 the Industrial Revolution began. The steam engine is given credit as a major catalyst. Prior to its invention, humans had harnessed water power (falling water) to turn wheels to power heavy machinery. That meant that activities were determined by the locations of moving water. Early steam engines burned coal, which has a high energy content-per-unit volume and was easy to use. With machines able to provide useful energy on-demand, manufacturing began in earnest. An early application of the steam engine was the railroads, which allowed large amounts of materials, livestock, and humans to be moved over long distances after railroad tracks were installed.

An**other major milestone was the invention of the internal combustion engine**. The first commercially successful engine was created by Étienne Lenoir around 1860, and the first modern internal combustion engine was created in 1876 by Nicolaus Otto. From his invention sprung the automobile and heavy-duty farm tractors and other agricultural equipment. **These latter machines didn't really materialize until oil came along.** That began after Edwin Drake drilled the first U.S. oil well in Western Pennsylvania in 1859. Prior to oil, kerosene was widely used for lighting; it came from coal, which gave way to kerosene derived from oil. Other products emerged from oil over time, to the point we are at today.

Drake (right) in front of the well

Two other inventions deserve special attention in this short summary. The first is radio, developed by Marconi in the 1890s. It allowed humans to receive audio information at vastly more locations than the earlier signal-sending device, known as the telegraph, which required wires to connect the sender and the receiver. **Radio shrunk the world by providing wide-scale communications between people**. During World War II, radio was used to advantage by both sides of the conflict.

The other notable invention was television. It was conceived in 1927 by Philo T. Farnsworth, a farm-boy genius, while he was in high school. He was inspired by his plowing fields back and forth (early television involved rastering electron beams back and forth at extremely rapid rates). Television greatly expanded communication and entertainment worldwide. It evolved from Phil's early, bulky large vacuum tubes to today's thin screen devices that are part of our cell phones and modern home TVs.

To me mechanization and television are particularly "close to home." First, in the late 1930s and 1940 our family periodically drove from our home in northern Illinois to far north Wisconsin to stay with my grandparents on their very basic farm (outdoor "plumbing" (outhouse), an ice box that a huge man periodically brought a large chunk of ice to cool, and water from a pump on top of a well in the front yard). My grandparents were not well-to-do. My grandfather couldn't afford a tractor. The picture I'll never forget is grandfather guiding a plow pulled by a horse with the horse's reins around his neck, plowing one of his fields. Pretty basic. Not that long ago for me.

When I was very young, our house was heated initially by a coal-fired furnace. Coal chunks were dropped through a basement window into a coal pile next to our furnace. Later, fuel oil became available, and a fuel oil tank was installed to replace our coal pile. Finally, natural gas was piped into the neighborhood, and we had clean energy to heat our home.

The other experience was after I received my doctorate in nuclear engineering and plasma physics, I went to work with Phil Farnsworth in 1964. Phil had invented a unique method of producing nuclear fusion, and I joined him to contribute. Phil was a gentle, soft-spoken researcher from whom I learned a great deal in our few years together. One of his remarks in 1965 that I'll never forget: "If I'd known what they were going to do with television, I might not have invented it!" And that was the relatively bland programming of those days!

Back to modern fuels and by-products from oil. Without a doubt, we humans have benefitted immensely from abundant, low-cost energy in its many forms. In so many ways it's been an incredible boon. However, with human-induced climate change now a concern, we have a problem that that abundance has brought, and changes must be made. In addition, with oil being so important to us, and it being in finite supply, our oil bonanza is almost certainly in peril, as we will discuss.

Chapter V. Climate Change Enlightenment & Stupidity

A. Introduction

Concerns about human contributions to world climate change are rampant. Individuals, organizations, and governments believe that urgent actions are required to save mankind from a future climate disaster, specifically large temperature increases worldwide by the end of the century. The physics and chemistry of climate science are extremely complicated with many uncertainties remaining. In the following I attempt to simplify.

First, weather is not climate. Most people don't differentiate. Climate is weather averaged over many years, often a decade. Thus, the effects of an extreme hurricane one year may be washed out when averaged over many years of calm weather. Or a huge iceberg breaking off Antarctica one year but not followed by more in future years might be washed out of climate considerations.

Second, there are two kinds of climate change: natural and human-induced. Independent of humans, climate is always changing, which greatly complicates detailed analysis and forecasting. With respect to natural climate change, you might remember that the world suffered through the so-called Little Ice Age, from the 1300s to the mid 1800s. Misery was widespread, crops failed, people and animals died, etc. Wikipedia tells us "several causes have been proposed: cyclical lows in solar radiation (the sun dimmed), heightened volcanic activity, changes in the ocean circulation, variations in Earth's orbit and axial tilt (orbital forcing), inherent variability in global climate, and decreases in the human population … from the Black Death and the epidemics emerging in the Americas upon European contact."

It's estimated that the Northern Hemisphere cooled by less than two degrees Fahrenheit relative to temperatures in the 1950s, which we'll see was a pivotal decade. So, part of the earth's warming since the 1300s was natural.

For those readers who love graphs, NASA provides us global temperature variations since 1880, Figure V-1. There's nothing special about the selection of the years shown; it's just what they provided. The data are the little "o's" and Lowness is just a method of smoothing the data.

The temperature is seen to have declined almost a half a degree Centigrade (roughly a degree Fahrenheit) before it increased roughly a degree Centigrade (near two degrees Fahrenheit) today. **Climatologists believe that human impacts on the climate began after World War II, which would be roughly the zero level in the graph.**

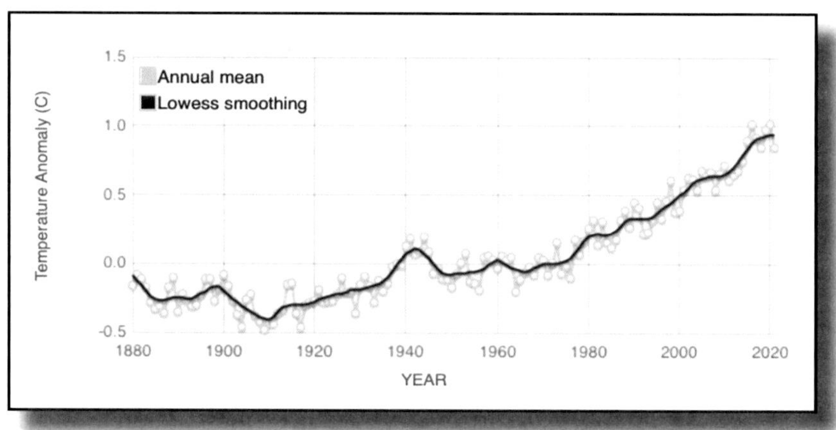

Figure V-1. NASA global land-ocean temperature index since 1880.

There is no question that climate change mitigation is necessary, but I believe that some of what's being done is stupid. The following provides: 1) a simplified description of the climate system; 2) the important greenhouse gases; 3) the emissions by economic sector; 4) the size of the emissions; 5) how imminent the problem is; 6) the IPCC omissions; 7) free world stupidity; 8) when do-gooders do dumb; 9) adaptation; 10) the geoengineering solution; and 11) your personal climate decisions.

B. The Climate System

The sun heats the earth with solar energy, which makes the earth habitable. About 30 percent of the solar energy that reaches our planet is reflected back to space and approximately 70 percent passes through to the earth's surface, where it is absorbed by the land, oceans, and the atmosphere, heating our planet. For various periods of time the absorption and reflection processes were in relative balance, and the earth's surface temperature stayed relatively constant.

Readers understand balance and imbalance – standing up versus falling down, bank account positive versus being short, etc. If more than a normal amount of solar energy is reflected away, the earth would cool, an imbalance. On the other hand, if more than a normal amount of radiation were retained in the earth-atmosphere system, the earth would warm, an opposite imbalance. It's the latter that's the climate change concern; **as part of our rapidly developing economies, we have emitted increasing quantities of so-called greenhouse gases that have caused the earth to retain more incoming energy, so it has slowly warmed.** That process has been called global warming, but more recently the more inclusive term "climate change' has been adopted to encompass more than just warming, e.g., more hurricanes, rising sea level, breaking icebergs, etc.

Collectively, even though some of the gases we've emitted only occur in trace amounts, they're believed to be sufficient to impact the climate, as we shall see.

The most important greenhouse gas is water vapor, known as humidity in our everyday lives. Water vapor accounts for between 35% and 65% of greenhouse gas emissions on clear days and between 65% and 85% when clouds are present. Part of the natural water vapor cycle involves humidity building up, after which it rises into the colder air that exists at higher altitudes, where it condenses forming clouds. Clouds reflect some of the incoming solar radiation (white surfaces reflect light), cooling the regions where the clouds are present. When the moisture in the clouds builds up high enough, water droplets form and fall to the ground….it rains. Clouds and rain occur over relatively small areas of the earth's surface; in other words rain is localized. Water vapor phenomena dominate climate but are not yet well characterized, because it's very difficult to get good information on humidity levels as a function of elevation above the ground and across the world.

Clouds, formed by condensed water vapor

Figure V-2 provides a simplified picture of the climate system. The figure comes from the IPCC, the Intergovernmental Panel On Climate Change, an international group of climate scientists and politicians who have periodically performed studies and issued reports on climate trends and climate forecasts since being formed in 1988.

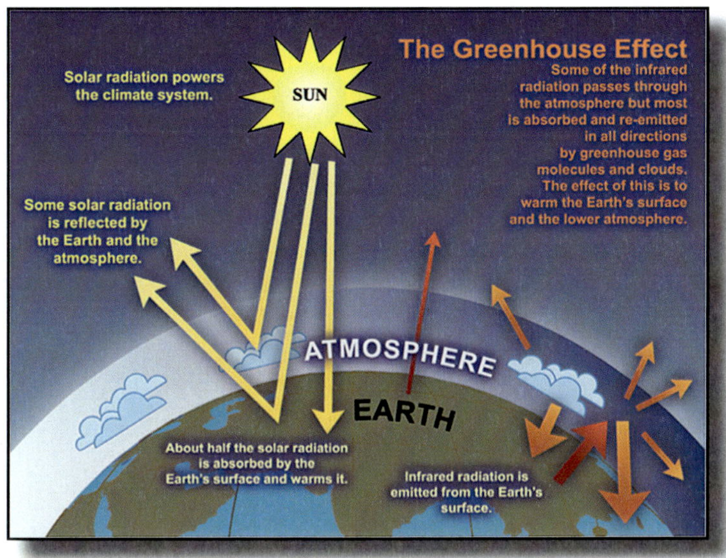

Figure V-2. A simplified picture of the natural greenhouse effect. IPCC. 2018.

Just in case that picture looks too simple, Figure V-3 provides more detail, which scientists must understand and model, if they're to be able to forecast what might happen in the future. **The whole system is unbelievably complicated and modeling requires advanced minds and advanced computers**. Climate models are continually evolving and provide the IPCC a means to estimate how the world might warm in the future.

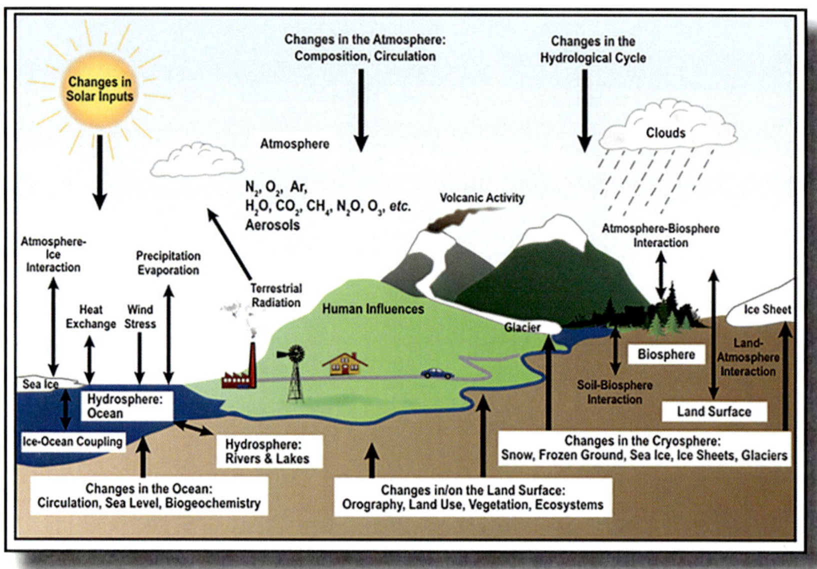

Figure V-3. Schematic of the components of the climate system from the IPCC in 2018.

To run their computer programs, modelers must make assumptions about future events out to the year 2100, such as how fast world economies might evolve; how human populations might expand; what volumes of gases are likely to be emitted each year; how technology might develop and impact people and the gases that are likely to be emitted; the extent of globalization; future agricultural patterns; the likely introduction of clean and more efficient technologies; etc.

Sound complicated? It is. As the IPCC itself says. "... the future is inherently unpredictable and so views will differ as to which of the storylines and representative scenarios (that the IPCC must develop to run their models) could be more or less likely. **Therefore, the development of a single 'best guess' or 'business-as-usual' scenario is neither desirable nor possible."** Their statement differs from much of what you hear from the media.

Also, because it's an international endeavor with participants from a large number of countries, all must be heard and satisfied that their views are properly included in the IPCC analyses.

Not surprisingly, the IPCC ignores the possibility of wars; it would be unacceptable for an international group to forecast wars.

Another extremely important topic has been omitted from IPCC considerations. **Most important in my opinion is the peaking of world production of coal, oil and natural gas.** At some point each will reach maximum levels, after which their production will decline.

Various people have attempted to forecast the dates of maximum production of these resources, but so far, all have been wrong. We're talking about some experienced, very talented specialists. With so many faulty forecasts in the past, it's not hard to understand why few are willing to provide new forecasts. Like forecasting wars, forecasting the maximum world production of various fossil fuels would be unseemly for an international "above-ground" scientific body, but these events will happen and will have dramatic impacts on the world and the human-induced drivers of climate change.

C. The Greenhouse Gases

Humans routinely release a number of gases into the atmosphere. As mentioned, the one that we're all familiar with is water vapor, obvious when we exhale on very cold days. Another that we see are the white cloudy emissions from chimneys and smoke stacks associated with industrial facilities. Yes, it's usually just steam (water vapor), not some concoction of bad stuff. The Environmental Protection Agency (EPA) has ensured that to be the case in its efforts over the past 50 years.

The other gases that we release that can impact the climate are called greenhouse gases. Figure V-4 shows the ones of greatest concern and their worldwide relative release volumes. As already touched on, fossil fuels include oil, coal, and natural gas (methane). **When the fossil fuels are burned, carbon dioxide results.**

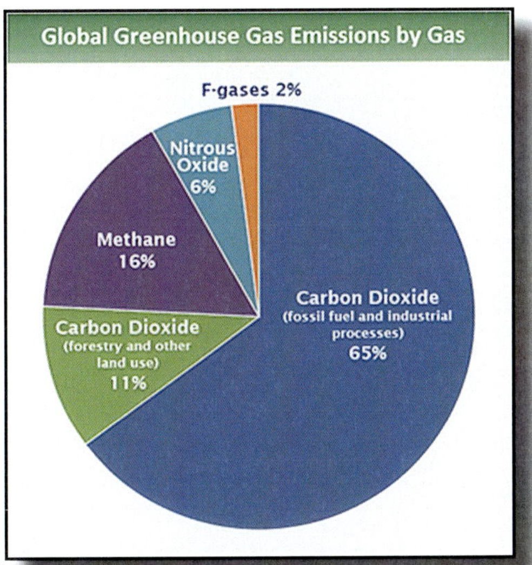

Figure V-4. Global greenhouse gas emissions from EPA.

Let's focus on carbon dioxide, which often gets a bad rap when discussing climate change. It plays an essential role in plant growth, called photosynthesis, which the dictionary describes as "the process by which green plants and some other organisms use sunlight to synthesize foods from carbon dioxide and water. Photosynthesis in plants generally involves the green pigment chlorophyll and generates oxygen as a byproduct." So more carbon dioxide in the atmosphere helps plants grow more rapidly, which is a good thing in terms of feeding humans and animals. Like anything, however, too much of a good thing thrusts us into the discussion of the levels that cause climate change.

The U.S. Environmental Protection Agency (EPA) gives us information about each of these emissions:

- "**Carbon dioxide (CO2)**: Fossil fuel use is the primary source of CO2. CO2 can also be emitted from direct human-induced impacts on forestry and other land use, such as through deforestation, land clearing for agriculture, and degradation of soils. Likewise, land can also remove CO2 from the atmosphere through reforestation, improvement of soils, and other activities.

- "**Methane (CH4)**: Agricultural activities, waste management, energy use, and biomass burning all contribute to CH4 emissions.

- "**Nitrous oxide (N2O)**: Agricultural activities, such as fertilizer use, are the primary source of N2O emissions. Fossil fuel combustion also generates N2O.

- "**Fluorinated gases (F-gases)**: Industrial processes, refrigeration, and the use of a variety of consumer products contribute to emissions of F-gases, which include hydrofluorocarbons (HFCs), perfluorocarbons (PFCs), and sulfur hexafluoride (SF6)."

Even if we were somehow able to magically stop our carbon dioxide emissions, there would still be important greenhouse gases remaining. **Climate change is not a single gas problem.**

There is no question that carbon dioxide has been building up in the atmosphere over recent decades. There are substantial arguments as to how much carbon dioxide is causing how much warming and how much climate change. Those matters are very complicated and discussions among specialists can be heated.

As a result of the buildup of carbon dioxide and other greenhouse gases in the atmosphere and stratosphere above it, **many believe that humans must rapidly replace oil/nat. gas and coal burning facilities with clean-energy technologies**. Accordingly, many countries are voluntarily shutting down existing fossil fuel power plants and replacing them with renewable energy generators, always at higher costs, because alternatives are more expensive, no matter what the proponents say. In addition, there is a major push to replace standard automobiles with electric cars, standard trucks with electric trucks, and diesel trains with electric trains, etc.

The bottom line is that we're dealing with a worldwide problem with countries other than the U.S. being responsible for the vast majority of the world's carbon dioxide production, as we will see. Thus, if the U.S. alone were able to somehow stop all of its fossil fuel use overnight, the problem of carbon dioxide buildup in the atmosphere would be only modestly impacted. But Europe is strongly motivated to unilaterally cut its carbon dioxide emissions, so even if the two entities were to instantly stop all carbon dioxide emissions (can't be done), it would only be a reduction of 24%, which could slow carbon dioxide buildup in the atmosphere but would not stop it.

D. Emissions by Economic Sector

If you're like many people, your eyes have begun to glass over at this point (or sooner), because you've seen more data than many of you can easily assimilate. So forgive me for making things worse. Why? Because if you're really going to seriously think about these matters, you need some more information.

Figure V-5 gives data from the EPA on where greenhouse gases are coming from worldwide as a function of economic sector, e.g., industry, transportation, agriculture, etc. Just to note, EPA got its data from an IPCC 2014 report. The data is generally representative of what occurred in 2019, the last year before Covid hit, which caused all kinds of dislocations in economies worldwide, as readers well know.

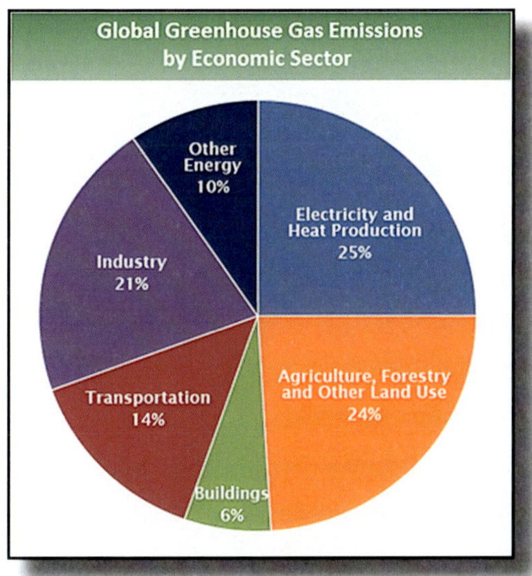

Figure V-5 Sources of greenhouse gases (EPA).

Briefly, here's what each includes, according to EPA/IPCC:

- "Electricity and Heat Production (25% of 2010 global greenhouse gas emissions): The burning of coal, natural gas, and oil for electricity and heat is the largest single source of global greenhouse gas emissions.
- "Industry (21% of 2010 global greenhouse gas emissions): Greenhouse gas emissions from industry primarily involve fossil fuels burned on site at facilities for

energy. This sector also includes emissions from chemical, metallurgical, and mineral transformation processes not associated with energy consumption and emissions from waste management activities.

- "Agriculture, Forestry, and Other Land Use (24% of 2010 global greenhouse gas emissions): Greenhouse gas emissions from this sector come mostly from agriculture (cultivation of crops and livestock) and deforestation. This estimate does not include the CO2 that ecosystems remove from the atmosphere by sequestering carbon in biomass, dead organic matter, and soils, which offset approximately 20% of emissions from this sector.

- "Transportation (14% of 2010 global greenhouse gas emissions): Greenhouse gas emissions from this sector primarily involve fossil fuels burned for road, rail, air, and marine transportation. Almost all (95%) of the world's transportation energy comes from petroleum-based fuels, largely gasoline and diesel.

- "Buildings (6% of 2010 global greenhouse gas emissions): Greenhouse gas emissions from this sector arise from onsite energy generation and burning fuels for heat in buildings or cooking in homes.

- "Other Energy (10% of 2010 global greenhouse gas emissions): This source of greenhouse gas emissions refers to all emissions from the energy sector which are not directly associated with electricity or heat production, such as fuel extraction, refining, processing, and transportation."

One of the sources that is often omitted from these charts is us. **The average human exhales over two pounds of carbon dioxide each day.** When you exercise vigorously, that number goes up significantly. Since there are close to eight billion of us on the planet, if we're all taking it relatively easy, that's 16 billion pounds of carbon dioxide we're adding to the air every day. That's roughly 8 million tons, which isn't all that much comparatively. Of course if you add elephants and horses..........

E. The Size of the Emissions

You need one more piece of information to put the foregoing in proper perspective. That's the magnitude of the carbon dioxide emissions. Consider just China, the world's largest emitter. In 2019 China emitted roughly 10 billion tons of carbon dioxide, 20,000,000,000,000 pounds. (Pounds are easier for most of us to grasp than tons).

A Boeing 747 weighs roughly 400,000 pounds, so the Chinese emissions are roughly equivalent to 50,000,000 747s. Viewed another way, the Empire State Building weighs roughly 360,000 tons, so the Chinese emissions are roughly 30,000 Empire State Buildings per year.

F. How Imminent is the Debacle?

A century ago people knew relatively little about climate science, so concerns varied dramatically. In the 1970s as more information became available, the worry was that we were on our way to significant global cooling. After all, the earth had gone through an Ice Age and a Medieval Cooling, so another cooling period seemed possible.

In the later 1980s and beyond, scientific understanding was based on ever improving solid information, and a widespread belief in impending global warming replaced the global cooling concern.

Confusion and political positions began to spread in the media with different sources making different forecasts, some suggesting dire conditions might begin relatively soon. We won't list the myriad of now-discredited climate forecasts. They simply showed that there was widespread uncertainty and political posturing. **To say that politics often impacts public discussions of climate change is a gross understatement.**

Thankfully, an international scientific effort was established (IPCC), bringing together some of the best climate scientists in the world, who have issued periodic reports on the status and outlook for the earth's climate. Their 2021 report provided a number of interesting findings, buried in incredible detail (Take a look for yourself!)

Recall, there are two types of climate change: Natural and Human-Made. As previously mentioned, the Ice Age and the Medieval Cooling were extreme climate changes caused by natural phenomena, well before humans had a chance to significantly make an impact. The Ice Age occurred roughly 2 billion years ago, during which huge ice sheets covered vast areas of the globe.

During the Little Ice Age, humans were around and kept records. Many scientists believe the Little Ice Age was caused by some combination of decreased summer solar radiation and/ or erupting volcanoes.

There have been a number of volcanic events over the centuries that were intense enough that the "stuff" the volcanos sent into the atmosphere cooled the planet. I'll address those events and the reduced global temperatures shortly

You might wonder how much heating of the planet has occurred thus far. NASA and the IPCC gave us Figure V-6, which clearly shows **the warming trend beginning after World War II, which is when economies worldwide were undergoing dramatic expansion and carbon dioxide emissions were increasing significantly**. The observed temperature increase between 1950 and 2020 is roughly a degree centigrade, which is 1.8 degrees Fahrenheit.

While the graph looks dramatic, the reader should note that the observed increase in 2020 is only 1°C (1.8° F). So, we humans have already accommodated to roughly a two-degree Fahrenheit temperature increase without serious complaints. As for animals, most seem to have accommodated, one exception being polar bears, which we're not sure about, in spite of heart-rending TV depictions.

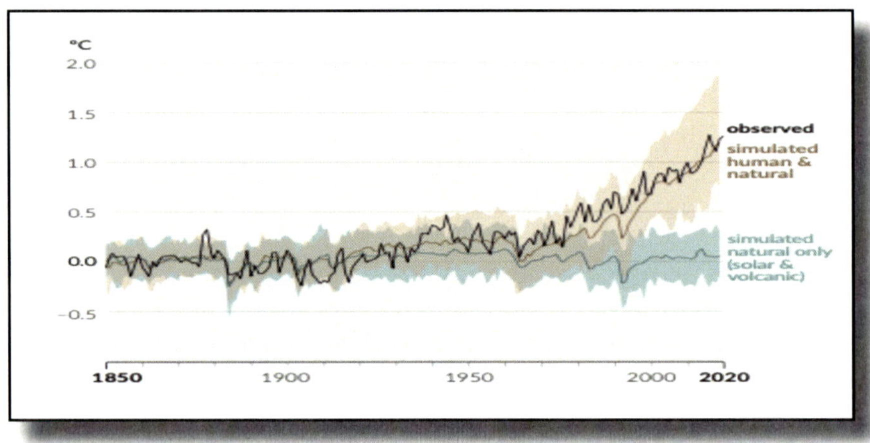

Figure V-6. Global temperatures since 1850, according to the IPCC

Finally, consider the recent IPCC forecasts to the end of this century, Figure V-7. The graph shows a number of cases running as high as an increase of five degrees centigrade to a low of just above where we are today. If you want detail on these cases, look at the most recent IPCC report. As arduous as it may be, I suggest you read the report itself, because the Summary for Policy Makers was written by politicians, not the scientists themselves. In fact, the scientists have been precluded from revising that Summary. Could policymakers be introducing exaggerations?

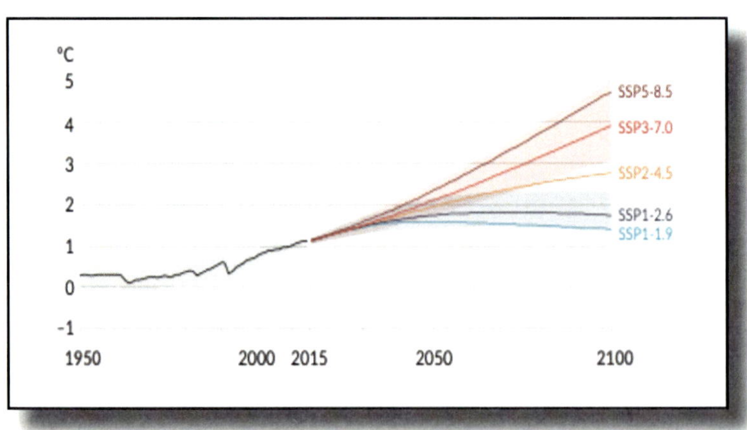

Figure V-7. The IPCC forecasts out to the year 2100.

G. The IPCC Omissions

As noted, it's virtually certain that world production of oil, natural gas, and coal will reach a maximum (peak) and then decline well before the end of the century. That will have a profound impact on economies worldwide and on their use, which will translate to dramatic changes in the emissions of carbon dioxide and other greenhouse gases. The IPCC didn't address this issue. I don't know if they even discussed it. Instead, the IPCC defined future world emissions cases that would bracket such a horrendous event. There are few major government organizations that would publicly discuss, let alone forecast the decline in world fossil fuel production, because it would almost certainly create widespread panic.

H. Free World Stupidity

Now consider the emissions of carbon dioxide by country. EPA gives us Figure V-8.

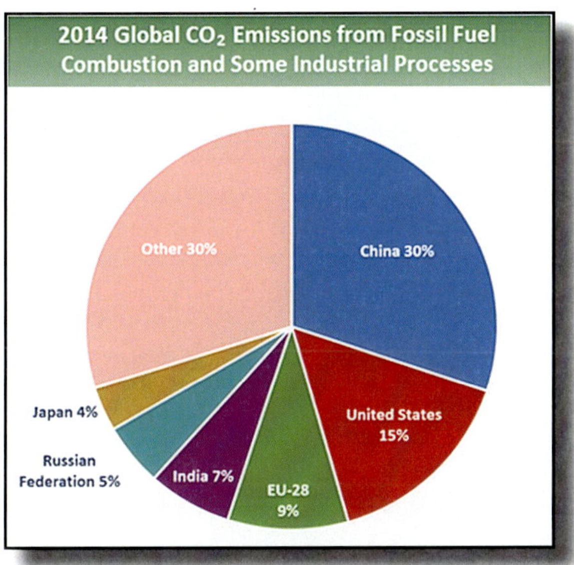

Figure V-8. The major world emitters of carbon dioxide from EPA.

The UN Framework Convention on Climate Change (UNFCCC) was adopted in 1992 and was one of the first international treaties on the topic. It stipulated that parties should meet regularly to address climate change. **In 2015 the United Nations (UN) organized the Paris Conference of Parties (COP) and adopted the Paris Agreement, which addressed climate change mitigation, adaptation, and finance.** The United Nations advertised it as a "legally binding international treaty on climate change," when it entered into force in 2016. "Its goal is to limit global warming to well below 2, preferably to 1.5 degrees Celsius, compared to pre-industrial levels." Sounds good, right?

But the U.N. has no enforcement powers, so realistically, the Agreement is just a piece of paper, which participants can honor or ignore. In fact, the U.S. did not ratify the Agreement. Instead, President Obama committed to it by executive order, because he believed he could not get it ratified by the U.S. Senate. Accordingly, the U.S. is not legally committed to the Paris Agreement.

It must be noted that the poor countries can't afford to ditch their lowest cost energy sources for more expensive ones. Whenever that problem is discussed, the poor countries blame the rich countries for climate change with reason, and the poor countries demand that the rich countries pay for them to change. That's of course a non-starter. Can you imagine a U.S. President going to the Congress and asking for $100 billion to a trillion dollars to give to the poor countries to cut their carbon dioxide emissions? It won't happen. Even if some moneys were to be sent to the poor countries, those moneys would almost certainly line the pockets of the dictators and their cronies. Not all countries are well run democracies!

In 2021 the U.N. hosted COP 26 to further stimulate climate change mitigation. Enhanced pledges were made and many agreed to work harder to mitigate a future climate nightmare. However, due to interventions by China and India, the conference ended up adopting a less-than-hoped-for resolution on ending subsidies for fossil fuels. The conference did commit to reducing coal use, but its use has reportedly increased worldwide since the conference.

Here's a matter that's rarely discussed: **A look at the globe makes it clear that some countries would benefit by global warming.** Why? Because those countries have large areas of land that are now near useless due to consistently low temperatures. A warmer climate would provide significant opportunities for agriculture on land that's not now useful for that purpose for instance. Russia, China, and Canada come to mind as countries that would benefit from global warming, but there are others.

So, we have an unenforceable U.N. agreement including countries that privately understand that they could benefit from global warming.

Finally, after Paris, China said it would seek to limit its carbon dioxide emissions before 2030, which is a change from the country's earlier commitment under the Paris Agreement to peak emissions "around 2030 and making best efforts to peak early." Even if it did make a commitment and didn't comply, there's no realistic way to enforce it or punish them.

Now for the stupidity of a number of countries, including the U.S. We started the 21st century using the least expensive energy sources, coal, oil, natural gas, and existing nuclear power. By our non-treaty agreement to commit to dramatically reduce our carbon emissions, **we effectively agreed to move to more expensive energy sources, which means increasing energy costs. That burdens millions of families with higher energy prices, and of course it impacts industries, some more than others.**

For energy-sensitive companies, higher energy costs means they become less competitive compared to similar companies in lower energy-cost countries. If higher energy prices appear to be a long-term situation, many of those companies might move to countries where energy costs are lower. Those moves mean lost jobs and increased national strategic vulnerability, when critical industries are involved. Some of that has already occurred, and because of 2020-2022 supply-chain disruptions due to Covid, the U.S. found itself disadvantaged with shortages of a number of products such as face masks and computer chips, to name just two.

The bottom line is that a number of countries are incentivized to stay with carbon dioxide emitting fuels to retain industries and attract others. Think China, which has been attempting and succeeding at luring critical industries from the U.S. and elsewhere, as part of its efforts to become the dominant world power. Think other countries that are trying to move ahead economically by sticking to lowest cost energy.

I. When Do-Gooders Do Dumb

Against this background, there are politicians forcing their regions to build expensive wind mills and solar cells to displace lower cost coal, oil, and natural gas-fueled facilities in the name of climate change mitigation. Why? They're motivated to do their part to save the planet. They want to "Do Good." You can't argue with their intentions.

However, look back at Figure V-8. What it shows is that **there's a big world out there producing carbon dioxide on a massive scale.** By shutting down a coal-burning electric power plant, a government is reducing world carbon dioxide emissions by such a tiny amount that it wouldn't show on the figure. In fact, shutting ten coal-burning power plants wouldn't show either. **It's like trying to drain Lake Michigan with a bucket. Those kinds of efforts are fruitless.** And they leave citizens with higher electric utility bills, not to mention that when electric power costs go up enough, it impacts industrial competitiveness and can drive companies out of the effected jurisdiction.

Consider what Germany did in the name of "doing good" on climate change mitigation. Germany is responsible for roughly 1.5% of world greenhouse gas emissions. That's very small but nevertheless Germany decided to dramatically reduce its emissions. In 2002 it adopted its National Strategy for Sustainable Development, making sustainability a guiding principle for national policies. As reported by Holman Jenkins in the Wall Street Journal, November 8, 2013, Germany provided subsidies to encourage homeowners and farmers to install solar panels and windmills and sell their excess electric power energy back to the power company at inflated prices. **By 2013 Germany received 25% of its power from renewables.**

Was this a climate mitigation success? As renewables capacity built up, carbon dioxide emissions surprisingly increased, rather than fell, because money-strapped utilities had to burn coal to provide the necessary standby power when wind and solar power waned. Wind and solar cells are intermittent, and people demand electric power-on-demand.

Because electric bills are paid by households and businesses, German electricity rates were triple those in the United States. Today, German electric bills are now the highest in the world. But Germany was the "industrial giant" of Europe. As one European official is purported to have said, "we're witnessing the 'deindustrialization of Germany.'" Do-Gooders doing dumb.

Look back at Figure V-8. **Even if the U.S. and the E.U. were to magically cut their carbon dioxide emissions to zero tomorrow, the result would only be a reduction of roughly 24% of carbon dioxide emissions in the world, which, all else being equal, would just modestly slow climate change.**

The climate change problem is a world problem.

Are we saying that there's no hope? Absolutely not! There are two distinct options: Adaptation and Geoengineering.

J. The Adaption Solution

One option for dealing with climate change is to do nothing and just let it happen. There's no indication that there will be a sudden awakening – a climate change Pearl Harbor - that will shock the world to advancing climate change. Climate change will happen gradually, as it already has, and people will either move further north or increase the size of their air conditioners. It's not inconceivable.

Furthermore, as previously indicated, the fossil fuels oil, coal, and natural gas will undoubtedly reach a maximum and decline before the end of the century, which means that the IPCC worse cases are unlikely. That means that **the likely climate change damage will almost certainly be much less than fossil-fuel-eliminators fear.**

By the way I've not forgotten the endangered polar bears and other animals that may not be able to evolve as fast as the climate changes. I share in related concerns. However, from a hard-ass point of view, people are the most important, and their survival should dominate the hierarchy of importance.

K. The Geoengineering Solution

The second mitigation option is geoengineering ... human-made technologies to immediately limit climate change. Volcanos "work" in cooling the planet. To quote the U.S. Geological Survey: "Yes, volcanoes can affect weather and the Earth's climate. Following the 1991 eruption of Mount Pinatubo in the Philippines, cooler than normal temperatures were recorded worldwide and brilliant sunsets and sunrises were attributed to this eruption that sent fine ash and gases high into the stratosphere, forming a large volcanic cloud that drifted around the world. The sulfur dioxide (SO_2) in this cloud -- about 22 million tons -- combined with water to form droplets of sulfuric acid, blocking some of the sunlight from reaching the Earth and thereby cooling temperatures in some regions by as much as 0.5 degrees Celsius. An eruption the size of Mount Pinatubo could affect the weather for several years." **Volcanoes have demonstrated one method by which the planet can be cooled.**

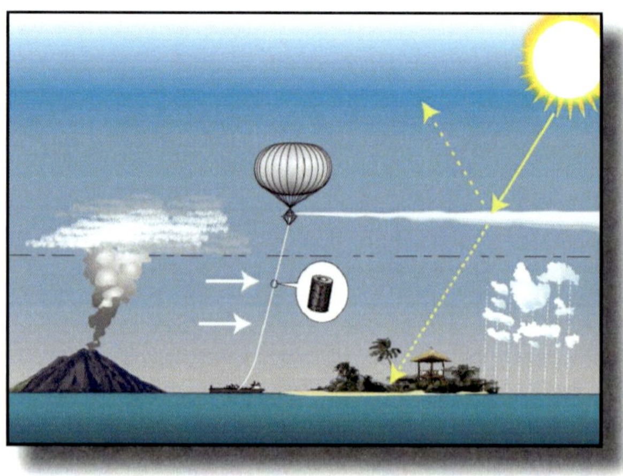

Proposed solar geoengineering using a tethered balloon to inject sulfate aerosols into the stratosphere.

Of course we have no control over when and where volcanoes erupt. However, we do have huge, reusable rockets to send large volumes of material into the upper atmosphere and stratosphere to mimic what volcanoes do, namely to increase the reflection of sunlight and cool the planet.

Human-controlled, volcano-like global cooling is one option in a field of technology known as geoengineering. It's a catch-all for various options to increase the reflection of incoming solar energy, thereby cooling the earth. Included are the general idea of adding small reflective aerosols or particles to the atmosphere/stratosphere; thinning high altitude clouds that can absorb heat; spraying sea salt particles to thicken ocean clouds; introducing large areas of aluminum foil in space to reflect sunlight; and growing more forests to absorb carbon dioxide to name but a few of the oft discussed options.

Some people think that they can efficiently remove carbon dioxide from the air. A few ventures are prototyping related ideas. I question this approach, because it requires a great deal of energy per ton of carbon dioxide removed from the air, and the problem is gigatons in size, which means the expenditure of energy from where?

Unfortunately, the U.S government has done little research on geoengineering, because factions in society want the world to stop use of all fossil fuels. The objections and lack of openness to other climate mitigation ideas smacks of an almost religious fanaticism.

Recently, the National Academies of Sciences, Engineering and Medicine were asked to do a study on geoengineering. Their March 2021 report recommends that a program be pursued in the U.S. "in coordination with other nations, subject to governance, and alongside a robust portfolio of climate mitigation and adaptation policies," I call this a "make nice, let's hold hands" kind of a program with all manner of U.S. agencies, environmental and public policy groups, and international involvement. Given that some people believe that it's likely a long time before geoengineering measures might be needed, this approach may be appropriate. On the other hand, many of us have been involved in such "make nice" programs over our careers and see such programs as magnets for all kinds of people and interests to slow, slow, slow down progress.

My preference is a program pursued by a get-things-done agency such as the Department of Defense, under a law that removes all requirements for public comment and environmental reviews, a "damn-the-torpedoes" approach. Such a program could quickly (a few years) get useful answers that could then guide more intelligent public policy. Getting quick geoengineering answers would minimize the time that well-meaning governments might use to hurt their citizens and industries. I believe this to be the superior approach to addressing climate change.

Of course we must be concerned that the environmental effects of any of the geoengineering options that emerge are potentially useful, environmentally acceptable, and cost-effective. In some cases we may not understand all related impacts until a technology is deployed at a significant level. It's fortunate that deployed geoengineering technologies can always be "turned off," if something objectionable were to be discovered.

One final point. Rough estimates of the cost of deploying effective geoengineering options indicate that the annual costs of deploying such options is likely measured in billions, maybe tens of billions of dollars per year. That's not insignificant, but it may end up as a bargain, when compared to having to refit most of our economy. Real costs won't be known until research and demonstration provide substance.

In summary, geoengineering is a near-term, relatively low-cost, minimally revolutionary solution to mitigating climate change.

L. Your Personal Global Climate Decisions

What might you do to help mitigate climate change? Are there purchase decisions you can make that will make a difference? The foregoing should have provided a clear answer. **There is nothing that you can do related to purchases that will impact world climate. Nothing. Nada**. Pragmatically, if you pay a premium for a product that purports to be good for the climate, when a similar not-so-advertised product is available at a lower price, you're not making a justified economic decision.

Right now, many state and local governments are making stupid decisions in the hope that they are contributing to climate mitigation.

You might also consider contributing to a non-profit committed to pragmatic action on climate change mitigation. The problem is identifying those that are pragmatic versus those that are using the climate issue as a means to raise funds for other interests. Good luck in such an endeavor; there are a number of wonderful-sounding appeals on the internet, radio, and television. Sorting through them will not be easy.

Finally, after reading our next discussion of the challenges associated with the sudden shortage of world oil supplies, you'll understand that oil shortage is your biggest pending energy threat. Accordingly, owning an electric vehicle will be advantageous when compared to owning a vehicle that runs on gasoline, which is derived from oil.

M. Conclusions on the Climate Change Challenge

The scientific side of the IPCC has clearly established that global climate change is a serious problem, which must be intelligently addressed. Because the summaries are written by politicians, I encourage you to read the report, rather than the summaries.

To this observer the current path to mitigating global climate change seems ho-hum; it's meetings and talk and meetings and talk. Some countries are mitigating alone while letting others off the hook. In the process they're doing economic damage to their own people, because they have and will lose industries to China and elsewhere, due to lower cost energy in those countries.

A more rational approach is required. I believe the best approach is geoengineering, because it offers the opportunity to take what appears to be economically modest actions, which could result in short-term climate mitigation under human control.

N. Post Script

As this book is being written, world oil supply is adequate to meet demand, although oil prices have recently been increasing to levels not seen for years. Banks and other significant organizations are forecasting that oil prices will continue to increase. More on that later.

One of the good things to come from the last few decades of climate change activities is the growing effort in the advanced economies to develop and deploy electric vehicles – cars, trucks, and buses.

In particular, a good deal of discussion centers on electric cars, which a number of manufacturers now have on the market. These vehicles don't use gasoline, thereby trending towards decreasing our oil dependence, albeit only modestly in the near-term.

Why is that? It's because the U.S. automobile vehicle fleet is enormous, almost 300 million cars, which have an average life of well over a decade. If we were to take the average value of each of those vehicles to be, say, $15,000, that's a capital investment approaching $5 trillion. Recall that the U.S. gross national product was roughly $21 trillion in 2019, the year before covid hit.

Furthermore, the average cost of a new vehicle today is well over $40,000, so if we had to somehow magically replace, say, half our existing cars overnight, it would cost roughly $6 trillion. These numbers are rounded for simplicity, because we're talking an impossibility.

After a decade of rapid growth, there are now roughly 10 million electric cars on the road worldwide, representing about 1% of the global car stock. This situation illustrates that dramatic change in the energy system does not happen quickly. Nevertheless, it's a start that can be built on in the future, not only for automobiles but for trucks and buses. The point is that electric vehicle development has given us a leg up on reducing our dependence on oil in this important element of transportation. Having said that, the challenges ahead are enormous, as will be described.

Chapter VI. Some Perspectives on Oil

A. Introduction

As indicated earlier, oil and its by-products are everywhere: cars running on gasoline, buses and trucks running on diesel fuel, airplanes running on jet fuel, roads made of asphalt, boats, ships, trains, lawnmowers, jet skis, motorcycles and a myriad of other machines.

In addition to fuels, oil provides detergents, plastics, lubricants, solvents, roofing materials, antifreeze, bandages, CD's, antiseptics, insecticides, pharmaceuticals, perfumes, refrigerants, tires, aspirin, shampoos, enamel, paints, dyes, insect repellant, trash bags, polyester, nylon, contact lenses, shaving creams, toothpaste, paints, soaps, epoxy, etc.

We shouldn't forget the construction of buildings, bridges, and homes for example. Building materials have to be transported to construction sites by gasoline or diesel-fueled trucks and trains, often over long distances. At construction sites, gasoline and diesel-fueled machines move construction materials into position, and the people doing the building get to work in vehicles primarily fueled by oil-based products. Modern civilization with all its wonders and flaws was founded on oil. We need to appreciate that fact and recognize that it won't always be available in needed quantities and at reasonable prices.

B. Oil Is Energy

It's essential to recognize that different machinery and equipment are built to operate on specific forms of energy. Computers run on electricity; obviously, you can't operate them on gasoline. All but a few automobiles run on gasoline; they can't be plugged into an electric outlet and powered by electricity, because they're not built to operate that way.

Electricity powers many of our appliances, computers, cell phones, subway trains, and a myriad of other devices. We plug them in to operate them; we don't pour liquid fuel into them and expect them to operate.

Essentially all liquid fuels are oil (petroleum)-based, meaning that they're composed of compounds of hydrogen and carbon -- hydrocarbons. That also means that when these fuels are burned with the oxygen in the air, they release carbon dioxide, which is a climate change stimulant.

Oil-derived liquid fuels are wonderful because of their high energy content, their convenience, their relative ease of handling, their relative safety, and their relatively low cost. Switching the energy we put into almost all machinery and equipment is impossible, because of the way the machinery is built. To run automobiles on electricity, for example, requires building a new type of automobile, which has happened in recent years. To operate heavy trucks on natural gas, for instance, we must either undertake an extensive retrofit of existing vehicles or build new ones.

Many long-distance trains operate on diesel fuel, and many short distance trains operate on electricity. To switch all trains to electric power, we would need to build new locomotives, add electric lines or third rails to train tracks, and construct new electric power plants to supply the needed power. These and other transformations can be done, but they'll take time and significant sums of money before national or world-scale change is achieved. These observations may seem obvious, but they're often ignored by people who claim that large changes in our energy system can be made quickly.

Many energy-form changes are attractive, but most existing-equipment owners are unable to discard what they have until the end of their equipment's useful life**. To think it easy or inexpensive to quickly switch energy type for a given piece of equipment is naïve.** Overnight machinery change is a fantasy. We can't afford to simply discard capital stock that has years of useful life left.

Because of this simple fact-of-life, the world's capital stock of oil product-fueled equipment will be with us for close to their natural lifetimes, unless a world-level shock changes the situation. **When dealing with oil supply problems, talk of windmills saving us is without merit. Windmills produce electricity; not liquid fuels**. Oil is energy, but not all energy is oil. Almost all equipment built to operate on oil products cannot operate on electric power.

Before discussing the perils ahead it's useful to understand the oil business in order to appreciate its vast size and why it works the way it does.

Chapter VII. Finding and Producing Oil

A. In the Beginning

Oil, often called petroleum or crude oil, is an incredibly complex mixture of hydrocarbon molecules composed of hydrogen and carbon. Oil was created millions of years ago by the geological burial of plant life, typically in ancient lakes, rivers and oceans. Over time, this plant life was buried ever deeper, where it was cooked by the earth's heat and compressed by heavy materials deposited over them over the eons. **In simple terms, lightly cooked plant material was coal; "medium" cooked material became oil, and longer-cooked material became natural gas.**

For the resulting oil, the end-result of these millions of years of burial and heating is an array of light, medium, and heavy oils. The variation in physical properties of oils depends on a wide range of complicated phenomena. In their natural form, crude oils are not useful in modern machinery, so considerable processing is required after oil is brought to the surface.

In simple terms, oil field activities involve the following steps:

1) ***Exploration*** to locate oil fields, which are not found everywhere. The largest are found in relatively few locations. An exploratory well drilled into a potentially productive formation provides the first real indication of the presence of oil and/or natural gas. Unproductive wells are called *dry holes*.

2) ***Appraisal*** follows a successful exploratory well and involves drilling additional wells to determine if the oil/gas reservoir is economic to produce. If little or no additional oil is found, the area is abandoned, which means that a large amount of time and money has produced nothing.

3) ***Development*** is the process where more wells are drilled and connected to pipes on the surface to connect wells and initial field processing facilities. Development often continues for decades. New wells are drilled and older wells are kept in good operating condition. Sometimes wells are re-drilled, because they can become clogged with sand and gunk. Onsite facilities must be maintained; wastes must be properly managed and disposed of, etc.

At some point the natural pressure in oil reservoirs becomes too weak to keep oil flowing at acceptable rates, so alternate means of pressuring or pumping are used. These often include injecting large amounts of water to push or float oil to producing wells. Sometimes enhanced oil recovery techniques are used to further produce oil that otherwise remains. These techniques might involve heating oil in the ground or injecting carbon dioxide to produce the oil that remains.

4) **Plug and Abandon** marks the end of an oil field. It involves 1) sealing the wells, usually by injecting cement to ensure isolation; 2) removing surface facilities; and 3) environmental cleanup.

Platform Harvest, offshore California

B. Oil Exploration

A simple quip from a friend: "Oil must be found before it can be produced." Finding oil has rarely been easy. The first U.S. oil well was drilled by Edwin Drake in 1859 in western Pennsylvania. The location was chosen because of blobs of oil that had seeped to the surface, suggesting that more oil might be available underground in that location. The drilling rig was very simple, but the venture established that oil could be produced in useful quantities, if one drilled in the right spots.

For a time the locations for oil drilling were determined by the presence of blobs of seeped oil on the ground surface. As time went on, acoustic techniques were developed whereby sound waves were sent into the ground. Those waves reflected off of different rock layers and were captured at the ground surface in surrounding areas on fancy electronic "ear phones." By analyzing the reflected sound waves, a picture of the subsurface could be developed. Over time, exploration people learned which reflected signals might represent oil-bearing rocks. Wells were drilled based on these interpretations. Sometimes they were successful and sometimes not – these were "dry holes."

Noble Energy sound wave generation (seismic) trucks

Over time explorationists (oil finders) and their research colleagues refined these "geophysical" processes until today they're highly sophisticated. In addition, petroleum geologists developed concepts for understanding how rocks were deposited and buried over time, and teams of both are involved in oil and natural gas finding. When exploration holes are drilled, the drill cuttings (chips) that are flushed back to the surface in the drilling process are examined to understand the rock layers that the drill bit has penetrated and to look for oil "shows." Occasionally, normal drilling is stopped and special drill assemblies are used to extract long "plugs" of rock called cores. These cores are examined to further develop a better understanding of local geology. When a significant number of wells have been drilled in a potentially interesting area, geologists will create what is called a "regional study" that shows expected geology over that area.

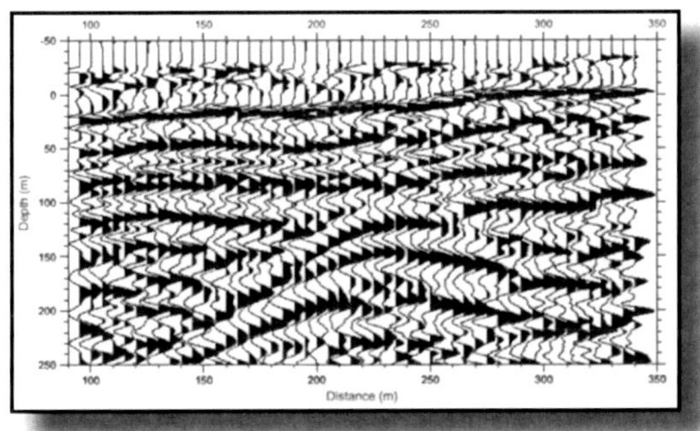

Exploration sound wave (seismic) reflection data

When I first became involved in exploration research, it was not easy for me to appreciate how very difficult exploration analysis was. Then I went on an exploration field trip with a number of ARCO oil finders. We traveled to a deep rock quarry but were not allowed to see it. First, we were given a number of widely spaced electrical traces of the quarry-face and asked to describe the rock formation from the traces. This mimicked what explorationists do in new areas of possible exploration interest after they've acquired information from a few wells that didn't produce any oil or gas. Of course those areas don't have deep quarries to provide clues about the rock formations.

I was a novice, but I tried to envision and draw the likely rock formations as best I could. The others were seasoned, successful oil finders. After we had all drawn our best guesses for the rock formations, we were driven to the quarry to see the actual formations. All of our pictorials were wrong! That's how tough the exploration problem is.

You can gain some insight into subsurface geological complexity for yourself. Next time you drive through mountains where there are road cuts, look at the exposed rock formations. You may see rock layers at different angles, layers of rock pinched off, rock displacements next to fractures, fractures that are cemented closed, etc. The complexity that you see gives you an idea of what can exist deep below the earth's surface. It's not orderly and layered, as some might think. It's often a mess of complexity.

Hopefully, this discussion didn't put you to sleep. The purpose was to demonstrate that exploration for oil and natural gas is very complicated. **When the technology and personnel**

got really good in recent decades, explorationists might end up with maybe one successful well out of ten tries in a new exploration area. I've dealt with these people and can tell you they're rugged, sophisticated, dedicated people who fail more often than they succeed. That's the character of oil finding.

One takeaway**: It's time-consuming and complicated to find oil and natural gas. Without considerable investment up front, it doesn't happen**. This point is critical in our later discussion of the possibility of an investment-related maximum in world oil production.

Oil and natural gas are NOT found everywhere. Where they exist, both are found in reservoirs -- containers that held them over millions of years, inhibiting them from massively leaking up to the earth's surface. As indicated, some leak a little but not so much as to empty the deposit. Oil and natural gas are often found together, because the geological "cooking" can be variable.

Oil fields consist of one or more oil reservoirs. An oil reservoir is a deeply buried rock formation composed of porous rock that holds oil and gas in its pores; oil and gas are usually found together. **Oil and natural gas reservoirs aren't underground cavities; they're porous rocks with oil and gas in the pore spaces**. These "porous rocks are covered by impermeable (leakproof) cap rocks that inhibit oil and gas from migrating up and out.

We've all looked at building bricks and concrete construction blocks. Those blocks are like the rocks containing oil in oil reservoirs. How oil can flow through the associated tiny pores is really a marvel.

One more thing. The rocks in which oil and natural gas form are called source rocks. Most of the oil and/or natural gas created in them doesn't stay in the source rock. Rather, it migrates out to a nearby rock formation that will hold it under an impermeable (leakproof) cap rock – a reservoir.

Why do you need this information when most of the world's oil/natural gas comes from these cap-rock reservoirs? Because oil companies have recently found a way to produce the oil and natural gas that remains in these source rocks and super-tight shale rocks. That process is called production from shale. **To recover shale oil and gas is much harder, much less productive, and much more expensive than from conventional oil/gas reservoirs.**

The process used to produce from shales involves drilling down to the rock formation, turning the drilling bit ninety degrees, drilling horizontally out maybe miles, making holes in the horizontal drill pipe, pressurizing thereby fracturing (breaking) the rock along the horizontal section, filling the fractures with sand to keep them open. The resulting open fractures provide a short avenue for adjacent oil and/or gas to migrate into the horizontal wellbore and flow to the surface. The so-called shale oil opportunity opened up a new avenue for production, but, as we'll see, these complicated wells don't produce all that much oil for the effort involved.

By the way, I forgot to tell you about the first people on the scene after explorationists decide that an area might be interesting for oil and/or natural gas exploration. The new areas of interest might be in the U.S. or elsewhere in the world. Explorations obviously can't just waltz in and begin to explore. They have to have permission, which is not simple. Enter the land man. These are people who must contact the landowner or offshore owner to not only get permission to explore, but also to establish contractual terms that permit production over many decades, if potentially profitable oil/natural gas is discovered.

This whole process is extremely expensive, particularly since so few exploration ventures are successful. A great deal of investment is required up front to finance those efforts. **The term "wildcat" is applied to describe ventures in new geographical locations, where there is no history of oil/natural gas production. Related investors must be willing to take the very high risks inherent to these efforts.**

In situations where a government entity makes government-owned land or offshore regions available for exploration, the land man works with lawyers to establish that the contract terms associated with the sale are acceptable.

C. Producing Oil

People don't typically comprehend the size of world oil consumption, and the fact that it takes a long time to build major new facilities. **Many of us are conditioned by what's happened in computers and electronics, where remarkable changes in capabilities and costs have been implemented over the span of months and years**. Those rapid changes are due to the fact that electronics and microcircuits are inherently very small. "Small" allows for changes to be made and implemented quickly.

Energy facilities are inherently very large. Mega's (millions) and giga's (billions) abound. For instance, world oil consumption is roughly 100 million barrels of oil day. On an annual basis that's over 35 giga barrels – 35,000,000,000 barrels. **If we were to lay oil barrels end-to-end, a day's world oil consumption in barrels would circle the world more than twice. A year's worth of oil barrels would circle the globe about 700 times**. Suffice to say, BIG!

More perspective. Consider an oil field with 10 billion barrels of reserves. Again, this sounds big. However, once you recognize that many decades are required to extract that oil, it becomes easier to understand that at maximum production, a 10 billion-barrel oil field might produce roughly a million barrels per day during its best years, after which it will produce less and less in each ensuing year. **Thus, during its best years, a 10-billion-barrel oil field might satisfy roughly 1% of world oil demand.**

You should also recognize that huge oil fields take a long time to bring into full operation, because they're such massive, expensive undertakings. Consider, for example the Lula giant oil field, roughly 160 miles offshore Brazil, discovered in 2006. It's considered to be the largest oil field discovered in the last 30 years in the western hemisphere. It's estimated to have reserves of roughly eight billion barrels with a significant amount of natural gas. Located in water that's 7000 feet deep, the reservoir is another 6,600 below the bottom of the sea. It's producing roughly 100,000 barrels of oil per day now. Why so long? It's not easy to operate in what can sometimes be raging seas. In addition, it took time to build pipelines to the mainland to move the oil and gas to refineries. Estimates of the costs to bring the field to full productions are of the order of $50 billion dollars. Oil producers wouldn't be working in such extreme conditions if there were many good opportunities on land. **We're closing in on the end of the easy, big opportunities.**

Why am I boring you with this detail? To illustrate how involved, time consuming, and expensive the process of development really is, even after a giant oil field is discovered.

D. Understanding Oil Fields

Okay, you've now got a rough picture of oil reservoirs. But it's still more complicated, not like pouring water out of a pitcher or draining a bathtub.

An oil field can consist of one or more oil reservoirs, maybe layered atop each other or adjacent to one another. A bowl of water wants to drain downward but is contained by the bowl. In an oil reservoir, the oil wants to escape upward, due to various physical forces, but the cap rock acts like an inverted bowl to hold the oil in place. Like a bowl of water, an oil reservoir contains a fixed amount of oil, part of which we might be able to drain, when we drill oil wells into it, releasing oil to flow to the surface.

Remember, in an oil reservoir, oil is trapped in the rock pores, like water in a sponge. However, unlike a sponge, reservoir rock cannot be squeezed to release the oil. Rather, oil must flow through adjacent pores to find its way to oil wells, where it can flow out.

Adjacent oil reservoirs in an oil field are typically developed together for convenience, efficiency, and economics, because field development is such a complex, expensive endeavor.

A production profile of a giant oil field is shown in Figure VII-1, where the time horizon extends over many decades. What you're looking at is oil flow increasing as the oil field is developed. Often production reaches a maximum for a time (a plateau). This production level is determined by the capacity of the facilities that are built for the oil field. It rarely makes economic sense to over-build facilities; rather, their size is the result of careful analysis based on a slew of factors.

Water for your bathtub can be turned on full with a turn of a knob; an oil field cannot. Your bathtub can continue to flow at maximum capacity until you turn it off; oil fields don't work that way.

Beyond the plateau is the decline period, during which engineers do their best to maximize and extend the decline as long as possible. Finally, it doesn't make sense to produce the little bit that's left in the field. It's not much in volume and not economic. At that point, wells are sealed with cement and the surface is returned to as close to its original state as is reasonable or has been agreed to in advance. That's called the "plug and abandon" phase.

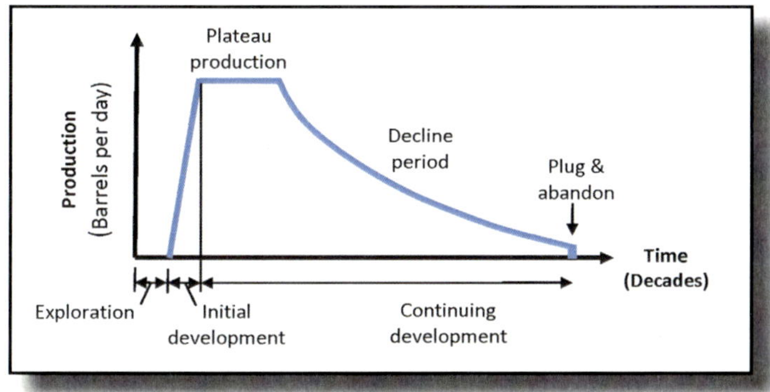

Figure VII-1. Oil production profile of a typical, large oil field.

After years of exploration and the drilling of successful appraisal wells, large-scale development begins. Oil production increases to a maximum and may stay flat on a production plateau for a number of years, sometimes a decade or more. Then, despite all human efforts, production decline sets in and can continue for many more decades.

There haven't been many oil field discoveries with upwards of ten billion barrels of oil (10 giga barrels). Figure VII-2 shows the record on giant oil field discoveries worldwide since 1850. **The world appears to have discovered most or maybe all of its giant oil fields and their related bonanza**.

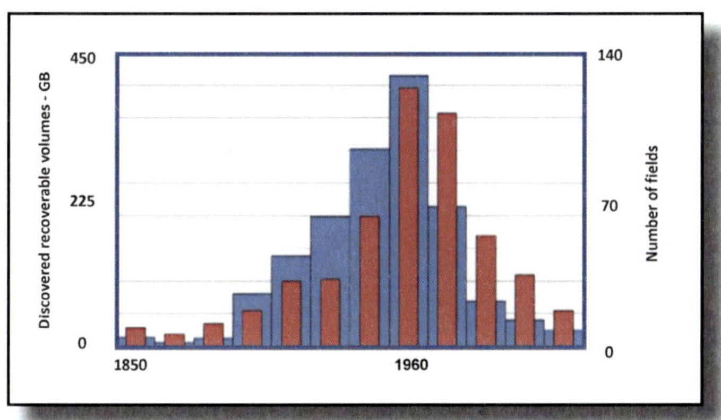

Figure VII-2 World giant oil field discoveries and recoverable oil. Blue is the volumes of the giants and red is the number of fields.

Figure VII-3 shows production profiles of four oil fields that are past their maximum production levels. As the figures illustrate, production profiles can vary considerably. For giant fields (Prudhoe Bay and Cantarell), a production plateau is often seen, while for smaller fields, the production profile can be relatively sharp.

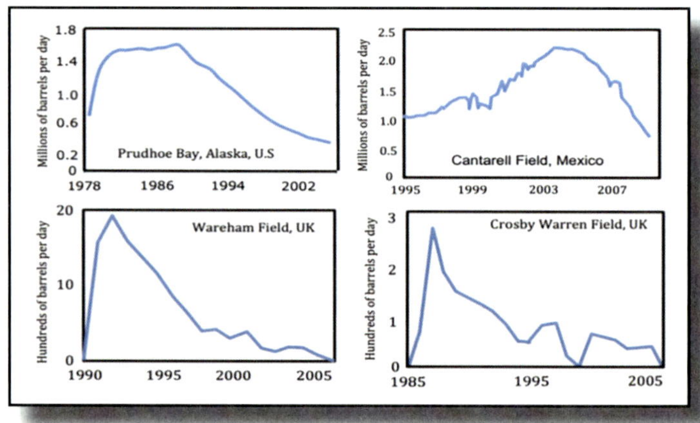

Figure VII-3. Oil production profiles for four oil fields of varying production outputs.

Plateaus are seen in larger fields, while smaller fields often show a sharp peak. Note the production levels: The giants are in millions of barrels per day while the dwarfs are in hundreds.

I know some of the history of the Prudhoe Bay oil field on the Alaskan North Slope. Atlantic Richfield (ARCO) and partners had taken a very large area lease near the Arctic Ocean. It's an incredibly desolate area. Explorationists drilled a number of wells on the lease and found nothing. I was told that ARCO and partners were ready to abandon the effort, but Harry Jameson, the chief ARCO explorationist on the job, pleaded for permission to drill one last well. Permission was granted, and the giant Prudhoe Bay oil field was discovered in 1968. It had roughly 25 billion barrels of oil, one of the very largest oil fields ever discovered.

One of the first oil wells in Prudhoe Bay in 1968.

The development of the Prudhoe Bay field was an enormous task, because winter temperatures range to 60 degrees below zero, and in the summer temperatures can reach the 60's degrees Fahrenheit. Summer days can range near 20 hours and are accompanied by mosquitoes that are the largest I've ever seen. In the winter the area has virtually no sunlight for roughly 60 days, November to January.

Besides having to invent ways to anchor facilities to the ground in this hostile environment (mushy tundra in the summer and hard frozen ground in the winter), a pipeline had to be built to carry the oil 800 miles from the North Slope to the port of Valdez in southern Alaska, where ocean-going oil tankers could take the oil to refineries in the Lower 48 states. That was another monumental task.

Peak production from the Prudhoe Bay field was 1.5 million barrels per day decades ago. Today, it's well into the decline phase, producing roughly 200,000 barrels per day.

Because the oil produced from Prudhoe Bay comes out of the ground hot, it can flow through the pipeline in Alaska's freezing winter all the way south without becoming frozen solid, which would be an investment and environmental disaster. It's interesting to note that the pipeline has a minimum flow below which the oil's thermal energy content would not be sufficient for the flowing oil to remain liquid on the trip south in the winter. That flow rate is not far below the current level. When that flow rate is reached, the pipeline and all production on the North Slope will have to be terminated, which will represent a non-trivial hit to total U.S. oil production.

The most important contributions to world oil production come from giant oil fields. They've been the easiest to find because they're big. However, we're not finding them much anymore, which doesn't bode well for future world oil supplies. The best data on global giant oil field discoveries goes back to a 2007 Uppsala University study and is shown in Figure VII-2.

I was coauthor on a 2009 analysis with colleagues at Uppsala on giant oil field declines. We found that "Their contribution to world oil production was over 60 % in 2005, with the 20 largest fields alone responsible for nearly 25%." Further, **"The overall production from giant fields is declining, because a majority of the largest giant fields are over 50 years old, and fewer and fewer new giants have been discovered since the decade of the 1960s."** In addition, "… this analysis showed that the average decline rate of the giant oil fields was increasing with time, reflecting the fact that more and more fields entered the decline phase and fewer and fewer new giant fields were being found." **So, production from the world's giant oil fields is declining, which bodes poorly for the future.**

E. Refining Oil

When economically viable oil is found, many more wells are drilled to provide multiple paths to extract (produce) it. Those wells are then connected by pipelines to facilities that do some initial processing, yielding oil that must then be sent by other pipelines to either a refinery, if it's relatively close by, or to tanker ships, which can transport the oil to more distant refineries.

Raw crude oil must be processed to produce the myriad of finished products we use in our everyday lives. That process occurs at a refinery, which usually covers a large area with a vast array of interconnected processing facilities that separate oil molecules into rough product streams, called "fractions," that are further refined to produce fuels, chemicals, plastics, pharmaceuticals, and other products. **Refinery processes perform functions such as molecular separations, composition changes, and impurity removal**. The final products must meet required specifications that provide optimum performance and environmental acceptability.

ExxonMobil oil refinery in Baton Rouge, Louisiana; the fourth-largest in the U.S.

After refining is complete, each product is transported to markets (for instance to distribution facilities and then to gasoline stations) or to yet other refineries for further processing into other products. A huge system has grown up over time to meet the ever-increasing needs of modern economies. As demands for higher product quality have increased, the refining industry has risen to the challenges, sometimes reluctantly. It's not that refiners don't want to provide cleaner products; they just don't want to spend money on new equipment, because refining profit margins are often very small.

As technologies advanced, oil exploration, production, transportation, refining, and distribution activities have become progressively safer and increasingly more environmentally friendly. Nevertheless, like any large-scale industrial activity, there are periodic accidents, spills, explosions, etc. In many cases these events were tragic and heart-rending when people, animals, or wildlife suffers. It's essential to remember that nothing is perfect – not any of us or almost any of the things we do. However, considering the huge volumes of hydrocarbons that are processed every minute of every day, the record is remarkably good.

Chapter VIII. Oil Field Reserves

A. Introduction

An estimate of the amount of oil that's producible over the life of an oil field is known as its reserves. Estimating reserves is difficult, because of unseen rock complexity, which is often better understood as more wells are drilled to develop an oil field. Accordingly, updated estimates of reserves are periodically made.

In some ways, oil reserves are like inventory, defined in the dictionary as *"The quantity of goods and materials in stock."* In the case of boxes of merchandise in a storeroom or liquids in a tank, inventory can be removed at a rate dictated by humans, e.g., ten boxes per day, so many gallons per hour, etc. However, the inventory analogy for oil fields is severely limited and can lead the non-expert to incorrect conclusions about oil field behavior. Figure VIII-1 is a simple-minded picture of the situation.

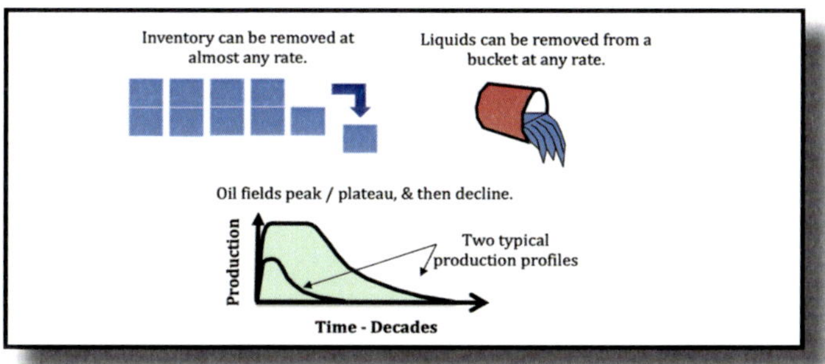

Figure VIII-1. Oil reserves are not like normal inventory.

Boxes in storage or liquids in a bucket can be withdrawn at a rate determined by people. After the onset of production decline in oil fields, the withdrawal rate is dictated primarily by nature.

People sometime think that large reserves automatically means a high level of production, which is not always the case. For instance, Venezuela has huge oil reserves, roughly 20% of the world's total, but decades ago the government began to view their oil production as a money machine to fund other priorities. In the process, the government starved its oil enterprise of the funds needed to maintain production. As a result, oil production deteriorated dramatically, as indicated in Figure VIII-2. The lesson: **If you mismanage oil fields and associated facilities, they deteriorate.**

You might think that competent management could move into Venezuela and restore or even increase production. It's not that simple, because Venezuela has exhibited decades of erratic and unreliable government, so developers hesitate to enter the country to undertake oil production because of the history of broken contracts and a hostile business environment. Potential outsiders must have confidence in a host country before they'll commit large investments.

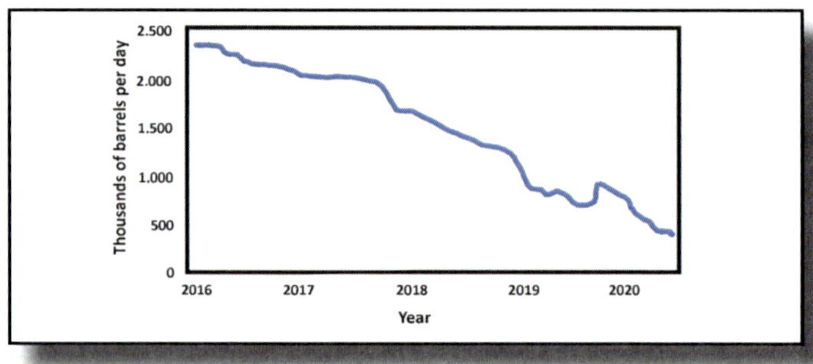

Figure VIII-2. Recent Venezuelan oil production, showing the drop off associated with gross mismanagement.

Estimates of oil reserves are dependent on the potential price of oil over the lifetime of the field. Why? Because when high oil prices are expected, an operator can justify drilling more wells and spending more on development. If low oil prices are forecast, less investment is justified, resulting in less ultimate oil production, meaning lower reserves. Reserves estimates are thus oil price sensitive.

Estimating possible future oil prices is a huge challenge. Going out 10 years or more? Consider how oil prices have fluctuated over recent decades, Figure VIII-3. Knowing that history, oil price prediction is no place for the timid!

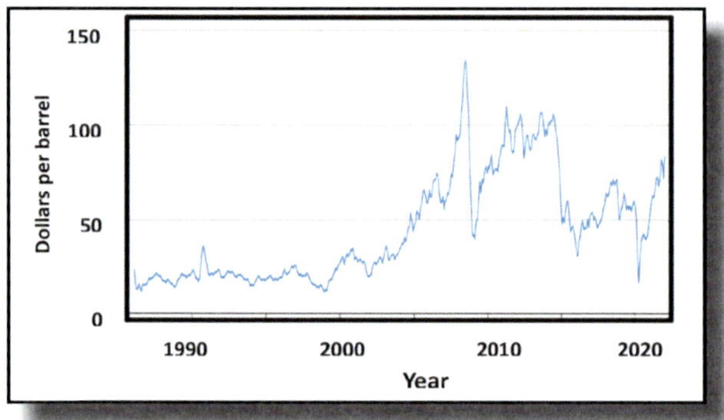

Figure VIII-3. The price history of U.S. oil over recent decades, as provided by the U.S. Energy Information Administration.

B. Oil Prices

The modern era of world oil pricing can be traced back to 1960 and the formation of OPEC, the Organization of Petroleum Exporting Countries. It was and is a cartel with the goal of controlling world oil prices. The initial five members were Iran, Iraq, Kuwait, Saudi Arabia, and Venezuela. Over time OPEC expanded, adding Algeria, Angola, Equatorial Guinea, Gabon, Libya, Nigeria, the Republic of Congo, and the United Arab Emirates. More recently, other countries, including major oil producer Russia "coordinate" with OPEC, which is now referred to as OPEC+.

Formally, OPEC's purpose was and is to "coordinate and unify the petroleum policies of its member countries and ensure the stabilization of oil markets, in order to secure an efficient, economic and regular supply of petroleum to consumers, a steady income to producers, and a fair return on capital for those investing in the petroleum industry."

Looking back at Figure VIII-3, which shows oil prices for the past 30 years, it's obvious that OPEC's price controlling efforts have not led to relatively stable oil prices. Some of the reasons include the following:

- OPEC has sometime been called a dysfunctional family, because the members are often in conflict, resulting in decisions that may not be optimum.
- The 2007-2008 global financial crisis caused many world economies to crater, dragging down world oil demand.
- Russia is a very large oil producer, and at the margin its high level of production has been a factor in various oil price declines. As of this writing, however, Russia has been ostracized from many marketplaces.
- The Covid epidemic of 2020 and beyond resulted in significant oil demand decline. It took time for economies to recover, stimulating oil demand with attendant oil price increases.

Finally, there's the advent of oil and gas production from shale, called by some as the U.S. shale revolution. Related production grew to the point of impacting U.S. and world oil production levels at the margin, meaning that in spite of its modest size compared to total world oil production, it upset the balance between supply and demand. Considering the size of world oil production, that's no small task. Figure VIII-4 shows U.S. oil production for over a century. **It's clear that conventional oil production (the inverted "V" pattern) was in decline until roughly 2010, when growing shale oil production began to impact on a major scale.** EIA recently said that more than 70% of U.S. oil production is now from shale.

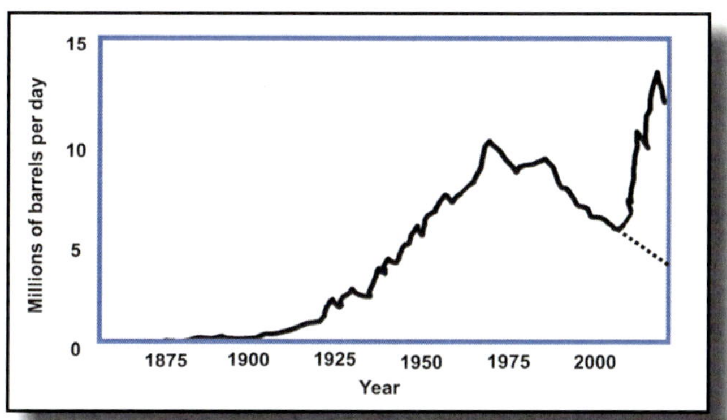

Figure VIII-4. U.S. oil production, showing the dramatic shale oil production uptick starting around 2010. The dotted line shows how conventional oil production continued to decline. The spike at the right is shale oil production.

We've long known that deeply buried shale rock contained oil and natural gas. The problem was, and is, that shale rock is so "tight" that simply drilling into it won't produce economically viable amounts of oil or gas. What do we mean by "tight"? It's the condition where the pores (openings) in the shale rock are so small that oil and gas cannot easily flow. To tap this resource, dramatic innovations in technology were needed.

Today's shale oil wells involve drilling a vertical well, changing the drill direction to horizontal, and continuing it, often for more than a mile, trying to follow the shale bedding. After drilling is complete, holes are made in the horizontal drill pipe and the pipe is pressurized, creating vertical fractures in the shale near the holes. Finally, those fractures are held open by flowing sand into them. Sound complicated? It is. I can't think of anything in everyday life to help the reader's understanding of this overall situation.

Finally, we note that producing oil shale is very expensive in part because the production of an oil shale well drops off in its first-year of production by a dramatic 50-80%. That compares with a typical vertical well into a conventional oil reservoir that might increase dramatically in its first year and stay relatively high for a few years before slowly declining.

So oil shale production involves many, many more wells than are needed in a conventional oil field to produce a comparable amount of oil over the longer term. Figure VIII-5 shows the production profile of a shale oil well, where the dramatic decline is measured in months, not years.

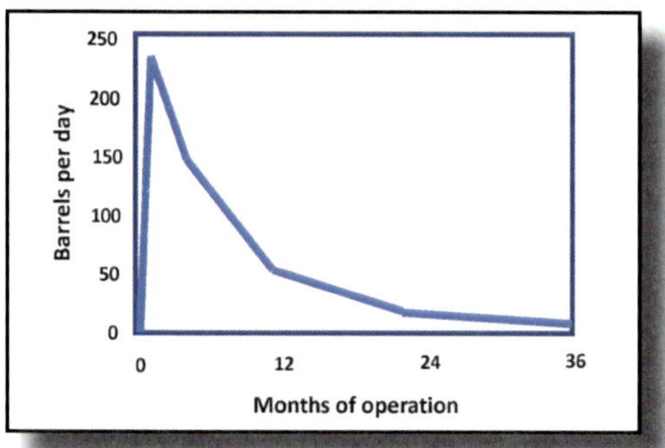

Figure VIII-5. Typical shale oil production profile, showing rapid rise followed by rapid decline.

As of 2022, the United States, Canada, China, and Argentina are the only countries actively producing commercial volumes of shale oil and natural gas, the U.S. being far and away the most active.

Shale oil is saving U.S. production, even though it's expensive and requires many, many more wells than would otherwise be the case with conventional oil. EIA tells us that the U.S. was using 20 million barrels of oil per day in 2019, the year before Covid upset our economy. In that year we imported 7.5 million barrels per day. The U.S. is heavily dependent on oil imports. **It was not and is not oil independent.**

C. Oil Reserves in the Real World

In the U.S. various oil company activities are highly regulated. Among regulator interests is the reserves a company can claim. Without regulations there might be an incentive for an oil company to claim larger reserves, so as to be able to attract more investment or influence its stock price. Enter the U.S. Securities and Exchange Commission (SEC), which monitors reserves estimates of U.S. companies, so we can generally take comfort that they'll be reasonable.

The situation in the Middle East is different. Consider the OPEC situation in the mid-1980s. OPEC decided to change its oil production quota formulas to take into account the reserves each of its members claimed. Higher reserves meant higher production quotas. Does this sound like a situation ripe for gaming? As described in Wikipedia:

"There are doubts about the reliability of official OPEC reserves estimates, which are not provided with any form of audit or verification that meet external reporting standards.

"Since a system of country production quotas was introduced in the 1980s, partly based on reserves levels, there have been dramatic increases in reported reserves among OPEC producers. In 1983, Kuwait increased its proven reserves from 67 Gbbl to 92 Gbbl. In 1985-86 the UAE almost tripled its reserves from 33 Gbbl to 97 Gbbl. Saudi Arabia raised its reported reserve number in 1988 by 50%. In 2001-02, Iran raised its proven reserves by some 30% to 130 Gbbl, which advanced it to second place in reserves and ahead of Iraq. Iran denied accusations of a political motive behind the readjustment, attributing the increase instead to a combination of new discoveries and improved recovery. No details were offered of how any of the upgrades were arrived at.

"This reality makes for real difficulties in calculating how much more oil is in the ground to be recovered, which makes the estimation of the date when world oil production goes into decline difficult to estimate."

Those dramatic increases in reserves estimates occurred during a period when relatively few new oil field discoveries were made in the OPEC countries, so the revisions were not based on new finds, nor was there a dramatic change in oil price. **The result is that the world was and still is faced with uncertainties regarding the amount of actual oil reserves in OPEC, which collectively has the largest reserves in the world.**

Furthermore, within most OPEC countries, oil reserves estimates are considered state secrets, and few countries have allowed outsiders to audit their published reserves. Some country estimates are obviously suspect. For example, since the "magic" jump in Saudi Arabian oil reserves in 1988, Saudi Arabia has stated that its reserves have fluctuated very close to 260 billion barrels for a number of years, in spite of the fact that Saudi production was over 3 billion barrels per year. In other words even though significant oil was produced, the oil that remained was claimed to remain constant. It's plausible that some reserves re-estimation and some new discoveries might balance production over a year or two or maybe three. However, Saudi Arabia has not had any major new oil field discoveries for a long time. Thus, to claim essentially constant reserves is asking us to believe something that's not plausible.

OPEC has long had an interest in the world thinking that there will be no world oil production limits in the future. Why? If their claim that huge supplies were accepted as true, importing countries would not be motivated to seek alternatives to imported oil, and OPEC could continue to sell its oil at stable or increasing prices. **Self- interest is not hard to understand, so OPEC's motivation should not be a surprise.**

Some have challenged Saudi oil claims. Others, who are conceivably in a position to do so, have not. Why? Consider some of the possible challengers and their circumstances:

- Many **oil service companies** provide oil field services to OPEC countries. As such, they may have insider information about oil realities in regions where they operate. They could raise questions about OPEC member claims, but that would involve biting the hand that feeds them.

- Some of the **major oil companies** formerly (or currently) operate in OPEC countries. Those companies and others hope for OPEC country business in the future, so it's not in their self-interest to challenge OPEC claims and get frozen out of future opportunities.

- The **U.S. government** (USG) has had a special relationship with the Saudis for decades. In past tight oil supply situations, it's rumored that the Saudis accommodated U.S. special requests by periodically providing relief. The USG recognizes that there might be more such situations in the future, so why would the USG risk antagonizing the Saudis by challenging their reserves claims? U.S. security organizations may have special insights into what the Saudis or OPEC actually has, but they're not made public.

- **Oil field consultants** could raise issues and some have. However, many consultants receive financial support from OPEC countries or might in the future, so they have little interest in antagonizing existing or potential clients.

- A number of **oil industry publications** could raise questions, but their business is selling their journals and magazines, so why risk losing lucrative markets?

- The **International Energy Agency (IEA)** could challenge OPEC, but if they did, they could lose sources of information, which could be to their disadvantage. About the most that the IEA has said is that IEA has no reason to doubt what OPEC has claimed.

So the world maintains major dependence on OPEC oil production with little substantive knowledge of what they really have. OPEC is in effect asking oil importers to "Trust Us." This is certainly a huge risk for the world.

Chapter IX. How Long Adequate Oil?

A. Introduction

Now that we've guided you through a simplified (!) picture of world oil basics plus some insights, let's turn to some history of supply-demand, more on reserves, and the outlook for future world oil production.

B. World Oil Supply and Demand

Thankfully, world oil demand and supply have been generally in balance, since the second world war. Figure IX-1 shows the situation since 1970, an arbitrary starting date. The supply and demand curves essentially overlap, so they're shown as a single line. Associated with that balance, explorers worldwide were seeking to find and develop new oil reserves with the result that reserves increased significantly, as shown in Figure IX-2.

There were two glaring exceptions to the supply/demand balance for the U.S. **1973 and 1979 saw Arab oil embargos of U.S. oil imports.** The result was widespread shortages of oil products. The situation was dramatic, with cars lined up at gas stations waiting, often in vain, for gasoline for their cars. Recall that in those days people drove to work and the grocery stores for food. The personal and national impacts were dramatically hurtful.

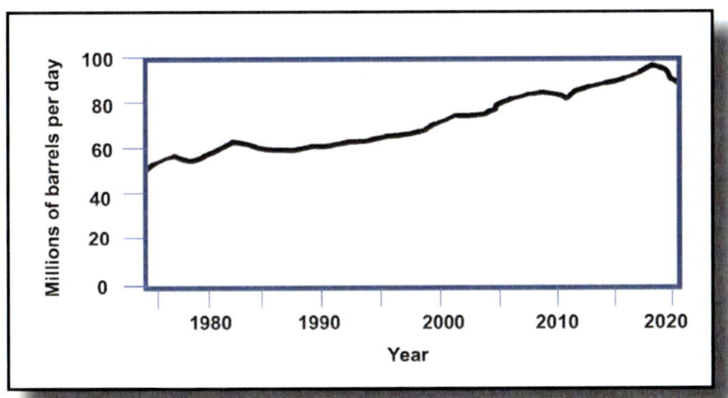

Figure IX-1. World oil supply/demand since 1970. Because this is the world situation, the Arab oil embargos of 1973 and 1979 are not evident.

Some of the studies used in my effort to provide you understanding are older than I might like, but the one's I use are believed to be directionally valid today. One notable analysis was done at Rice University in 2007. It stated that 77% of the world's proven oil reserves were controlled by the national (government owned) oil companies. International oil companies (IOCs like Chevron, Shell, etc.) controlled less than 10% with the remainder being joint ownership situations.

Table IX-1 shows the largest world oil companies by production, according to the journal Offshore Technology. The ones considered International Oil Companies (IOCs) are noted.

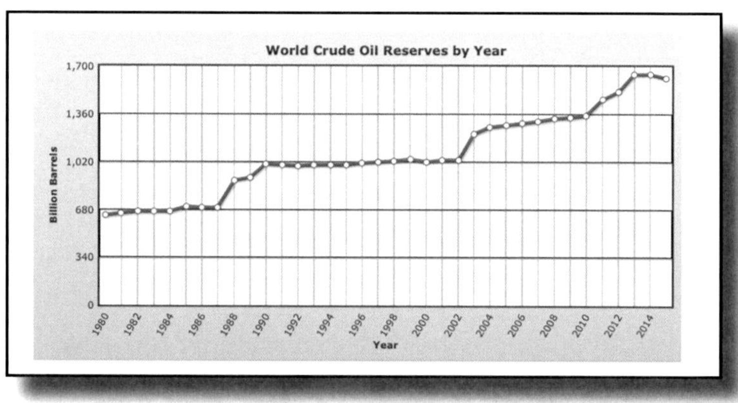

Figure IX-2. World crude oil reserves, provided by EIA. The rise in reserves is comforting for the future. The break around 2013 should not be read as significant until future data is acquired.

Table IX-1 World's Largest Companies by Oil Production

Saudi Aramco	**(Saudi Arabia)**	**National Oil Company**
Rosneft	**(Russia)**	**National Oil Company**
KP	**(Kuwait)**	**National Oil Company**
NIOC	**(Iran)**	**National Oil Company**
CNPC	**(China)**	**National Oil Company**
ExxonMobil	**(U.S.)**	**IOC**
Petrobras	**(Brazil)**	**National Oil Company**
ADNOC	**(United Arab Emirates)**	**National Oil Company**
Chevron	**(U.S.)**	**IOC**
Pemex	**(Mexico)**	**National Oil Company**

Why should we care? Because in a future world oil shortage, control of the supply will lie outside of U.S. control. At that time, hopefully, diplomacy will help the U.S. secure a fair share. If not, military incursions may be necessary. Does that sound too strong? If the U.S population is in chaos, extreme measures will be called for.

As already noted, the U.S. is not oil "independent." In 2019, the year before Covid hit and markets were skewed, the U.S. Energy Information Administration said we consumed an average 20.54 million barrels per day of oil of which 7+ million barrels per day were imports. So when world oil shortage occurs, the U.S. will be in a weak, oil import dependent position. From the foregoing discussions, you know that increased oil production cannot just be turned on quickly, especially in a country that's been extensively explored, like the U.S. .

[Where indicated, our numbers come directly from U.S. EIA. If you're keeping detailed track, you'll note some discrepancies. Welcome to some of the confusion an observer runs into related to these matters. One example of confusion: in some cases oil production includes the addition of 1) lease condensate – the modest amount of light liquids that are produced along with natural gas, 2) refinery gains – the liquid volume increase that comes from refinery processing, and 3) biofuels – ethanol from corn added to gasoline. These are usually secondary sources, so I've omitted them for simplicity.]

C. Maximum World Oil Production (Peak Oil)

At some point world oil production will reach a maximum and then go into decline, creating growing oil product shortages. No credible source denies this.

It's simple: Oil is a finite resource, which is being continually depleted. It's like a bucket that contains a finite volume of water; as you pour it out, you've got less and less in the bucket.

When will the maximum (peak) of world oil production occur? There have been numerous estimates of the year of the onset of world oil production decline by some very competent experts. The result: Failures.

Peak oil hasn't happened yet, even though a number of analysts forecasted that it would have. When experts are consistently wrong, they lose credibility, but it'll happen.

In the past, forecasters were estimating the year of oil production decline based on their understanding of the characteristics of existing and expected new oil fields yet-to-be-found. One key to past forecasts is knowledge of when the giant oil fields of the world were discovered and something about their production rates over the years. Reserves are important to know, but I've already indicated that the biggest world oil producers -- OPEC -- have not been truthful in the past about their oil reserves, and I've no reason to believe that they will all-of-a-sudden come clean. As a result, almost all forecasters were and are dealing with best-guesses based on limited information.

In the past forecasters based their thinking on geology, which was the primary basis for contemplating the peaking of world oil production. Recently, a large number of industry stalwarts have warned of a new kind of world oil supply shortage associated with the dramatic underinvestment in oil exploration and production over the last few of years.

So, we now have two kinds of drivers of world oil shortages: 1) Investment-Related, and 2) Geologically-Related.

At this point I won't bore you with the various opinions on when these two peaking events might occur; I've provided a number of them in an Appendix. **The bottom line is that investment-related oil peaking could occur in a matter of years.**

Geologically-Related peaking is still an obscure matter with the likelihood of it occurring in the latter 2020's or as late as the 2030's or 2040's. I know of no credible source that has a fact-based, published prediction.

Lastly, it could be that the onset of investment-related peaking will meld into the geologically-related peaking. Who knows?

Who might be a credible organization to report the onset of investment-related peaking? It's not likely to be an international oil company, because no oil company would want to take responsibility for creating the related world chaos, because chaos would surely ensue. If an IOC were to firmly believe that they knew the timing, they would surely first deliver their forecast to their host government, which would keep it as top secret, while it decided who to share the news with and how to prepare their citizens.

If our U.S. Energy Information Administration were to develop even a likely date for the investment-related peak, they would almost certainly classify the information and share it with the White House. How the White House would behave is unclear, but it's likely that the news would leak to the media in short order, because the government often leaks news, particularly huge news. If the International Energy Agency were the source, the news would be shared in secret with world governments and would almost certainly leak out quickly

D. How Are People Likely to Behave with the News?

Here we've some insights from what happened in the U.S. as a result to the sudden oil supply cutoffs of 1973 and 1979. In those two events, one of the first things that the public did was to rush to gas stations to top off the fuel tanks in their cars and fill their portable gasoline tanks for their power lawn mowers, boats, other appliances, and/or just to have some backup.

In 1973 and 1979 the gasoline distribution system wasn't prepared for such an onslaught, so gasoline immediately went into shortage, and gas stations quickly ran dry, leaving large numbers of motorists with no more than they already had. Back then, car pooling caught on quickly, because most people commuted to work. A related Appendix goes into detail about government-related actions that ensued.

One danger: Politicians did then and will again feel great pressure to do something to alleviate related problems. Under such pressure and likely without plans or a real understanding of how the oil world operates, they'll take actions that may not necessarily be in the country's best interests.

One obvious point: The U.S. of the 1970s is not the U.S. of today. Back then, people had to travel to their jobs and other activities. Today, after our experience with the Covid epidemic and with modern communications, a great deal of business can be performed electronically. Thus, the immediate impacts may be somewhat less severe than before. However, the period of a future oil shortage will almost certainly be much longer than the relatively brief events of 1973 and 1979, so the pain and suffering will likely be much worse.

Chapter X. Peak Oil Mitigation

A. Introduction

There are no "silver bullets" to quickly bail us out when world oil supplies become inadequate. The world's oil-based system is just too vast for rapid mitigation. Oddly, my technical career involved responsibilities in the federal government, industry, and the nonprofit sectors on a variety of technologies including renewables, conservation, oil and gas exploration and production, oil refining, synthetic fuels production, geothermal energy, nuclear energy, and fusion power. I can assure you that there is no perfect energy technology. All have limits and drawbacks. **Again, the issue under discussion is world liquid fuels shortages, not "energy."**

B. Liquid Fuel Options

1. Conservation

The first option in both everyday living and energy shortages is to conserve and carefully allocate what's available. Among other things, that means gasoline for most consumers will be curtailed. Air-travel will be severely curtailed. Distant vacations will likely no longer be viable, and many business activities can and will be performed electronically. Priority will be for the essentials of life – food. Farm vehicles and related infrastructure must receive fuels so they can remain viable. The military will also receive priority, because ready-to-act armed forces will be essential in time of peril. Various aspects of commerce will have to be prioritized and liquid fuels provided to the most important. Infrastructure activities will have to be prioritized, and many plans and projects terminated.

Oil refineries will change their product outputs to optimize for the most essential needs. For instance, more gasoline and less jet fuel. In some cases, dramatic refinery reconfigurations may be necessary so heavier, less essential oil constituents can be refined into priority fuels.

2. Maximizing Oil Exploration, Production, and Canada

Clearly, we'll want to produce as much oil as possible domestically, so priority will be given to maximizing production from any U.S. oil fields that are amenable. Oil that can be produced on the North Slope of Alaska and on federal lands will be given immediate clearance for development, after having been dramatically curtailed in the name of saving the climate. Human needs will be given priority over climate change and many other environmental concerns. That doesn't mean running roughshod, but it does mean ending time-consuming public comments and endless lawsuits before projects can get the go-ahead.

Canada has vast deposits of heavy oil, the production of which has been somewhat curtailed in the name of climate. Because Canada does not have refineries to process its heavy oil into useful products, its oil has been pipelined to the U.S. Gulf Coast for refining. Expanded pipelines will have to be built to handle much higher volumes. Part of the resultant products will be returned to meet Canada's needs, the remainder kept in the U.S. for our needs.

3. Liquefaction of Coal and Natural Gas

Both coal and natural gas can be converted to the refined liquid products that matches what oil production and refining provides. Coal-to-liquids plants have operated in South Africa for over 50 years. In the U.S., research and pilot plants were built in the 1970s and early 1980s to demonstrate the viability of coal and natural gas conversion to refined oil products. Because world oil supplies subsequently became abundant, those liquification concepts were never carried to commercial scale. The construction of commercial plants is possible, but they're large and expensive. Even with crash-program priority, the construction of such plants will take many years before meaningful quantities of liquids can be produced, and then even more will be required to catch up to meaningful production levels, as will be discussed shortly.

SASOL II and III, S. Africa. Coal-to-Liquids Facility.

4. Oil Shale

Shale oil was already discussed. **Oil shale is very different**. Huge deposits lie near the earth's surface and can be mined and processed to make important oil products. The richest oil shale deposits in the U.S. are in the Green River Formation located in Colorado, Wyoming, and Utah. Back in the 1970s, when the two oil embargos occurred, there was serious consideration of tapping the resource in Colorado, but there was one huge shortcoming that still exists today. Mining and processing requires large amounts of water, which does not exist where the oil shale is. There was some consideration of the long-distance piping of water into the region, but that would be a herculean task. Today, the water problem still exists, so tapping that resource would take a number of years, even with crash program priority.

5. Summary of Crash Program Mitigation

In 2005 coauthors and I prepared a study for the U.S. Department of Energy, entitled "Peaking of World Oil Production: Impacts, Mitigation and Risk Management." **The study developed scenarios for the very best that a worldwide crash program of oil shortage mitigation might accomplish.** Highlights of that report are provided in an Appendix. A brief summary follows.

First, we assumed two cases for the possible decline rates of world oil production: 2% and 4%. These are the rates that our crash program would have to race to mitigate. A “delayed wedge” method was used to approximate the scale and growth of each mitigation option, because that approach captured key elements of how things might work, and it simplified the analysis without loss of major insights. Figure X1 shows the delayed wedge pattern. The horizontal axis is time and the vertical axis is market penetration, measured in millions of barrels per day of savings or production. **A 20-year period was used because major mitigation impact inescapably requires a long time.** The delay before first impact is the period needed for preparation and/or facility construction before a technology begins to show noticeable results.

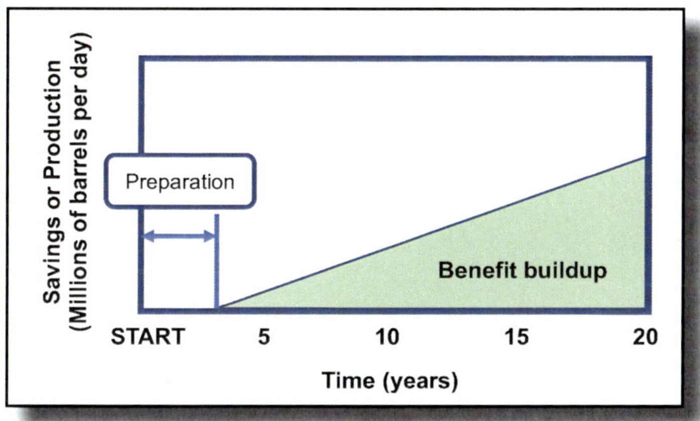

Figure X-1. Delayed wedge approximation for various mitigation options.

For the purposes of rough estimates of what worldwide crash program mitigation might achieve, the following were evaluated:

- Fuel efficient and electric transportation
- Heavy Oil/Oil sands
- Coal-to Liquids (CTL)
- Enhanced Oil Recovery (EOR)
- Gas-to-Liquids (GTL)

These same technologies would be appropriate today. The sum total of our results is shown in Figure X2, where our idealized, instant-start program is summarized. What you are seeing is roughly 30 million barrels of oil per day saved and/or produced in 20 years.

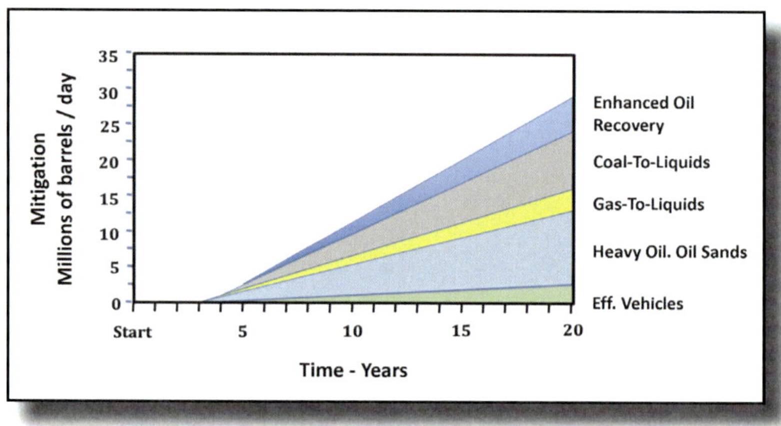

Figure X2. The sum total of crash program physical mitigation estimates.

Again, this worldwide crash program is racing against a decline of world oil (liquid fuel) supply. If we assume the program starts instantaneously, which is of course unrealistic in the real world, the question is how does this **best-case program** measure up against the two assumed world oil production decline rates. The results are show in Figures X3 and X4.

Pearl GTL in Qatar is the world's largest Gas-To-Liquids plant. (from Shell Global)

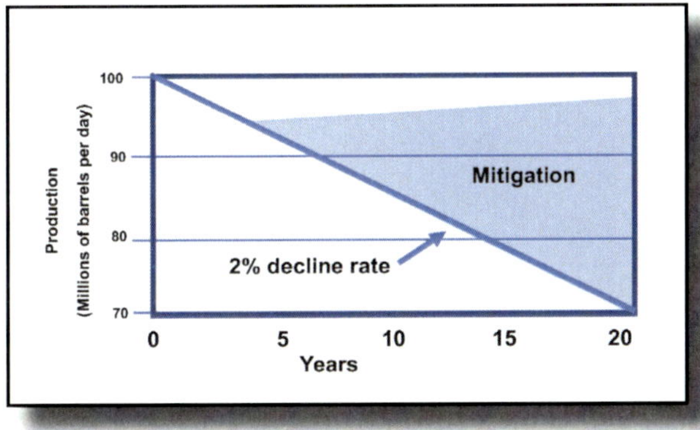

Figures X3. The impact of crash program mitigation on oil supply when world oil production declines at 2% per year.

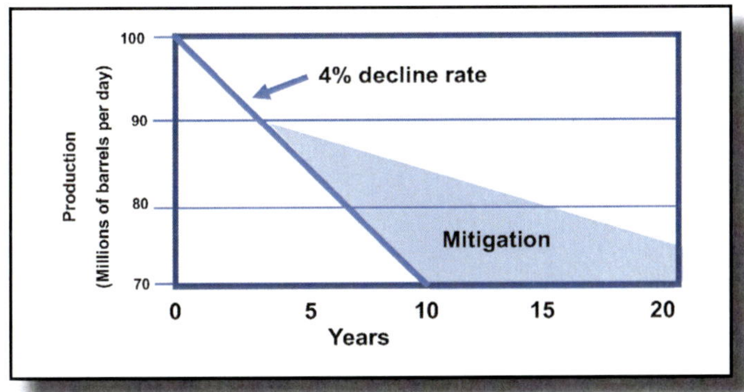

Figures X4. The impact of crash program mitigation on oil supply when world oil production declines at 4% per year.

The conclusions are that in the 2% world oil production decline rate, the overall production declines modestly and remains relatively flat at a lower level for more than 15 years, while in the 4% overall production case, the crash program mitigation never reaches a plateau. Unfortunately, no one knows what the world oil production decline rate will be after peak world oil production. This analysis of an idealized, monumental, worldwide crash program of mitigation was something never before attempted.

There's no question that an army of analysts could do a more detailed analysis. Such an exercise might be justified if the federal government was doing it a decade or more ahead of the advent of peak oil, as the government was preparing to mobilize resources in advance of the event. However, since the federal government is not noted for detailed long-range planning, it's hard to envision that happening at the full government level, but it could happen in the Department of Defense, if DoD was so motivated.

Chapter XI. Recommendations

A. Introduction

If indeed the world is facing one or both kinds of world oil shortages in the next decade or so, urgent action is required now, as the previous discussion indicates, because mitigation on the scale of world oil production will be time-consuming, even assuming a crash program approach. In the U.S. such an effort will necessarily be the responsibility of the federal government.

B. First, Establish the Likely Timing of Peak Oil

Obviously, something more than this book is needed to justify the effort that'll be required to mitigate the peak oil disaster. Planning and acting far ahead of the event is necessary, again, because **effective mitigation will require much more than a decade.**

A high-level analysis could be a precursor. Anything less than a Presidential Commission won't carry the weight needed to justify the enormous effort required to impact the problem. A timetable for analysis results and recommendations should be mandatory. My preference would be for a two to three-month study, which would be associated with the urgency the problem requires.

C. What Agency Would Best Take the Lead?

To me, the best hope for effective action is the Department of Defense (DoD). Why? Because they know how to make big things happen quickly and because part of the justification for the program would be to produce non-petroleum fuels for the DoD to provide for the national defense – think fuels and materials for planes, tanks, rockets, ships, and a myriad of other equipment. The Department of Energy would act in a critical supporting role.

The Congressional Act to authorize and provide funds for this effort should exclude the effort from external environmental reviews, because they're so time-consuming. That responsibility should lie with DoD. Sites for the requisite large projects should be a matter of DoD choice in consultation with other departments. The effort should be on a crash program basis, because normal-efforts would fall far short of the need. In the process, not only will facilities be built but issues of materials, technologies, and trained personnel will be defined and addressed.

D. The Push Backs

There will be vehement resistance to this approach by at least two groups:

The Middle East oil producers will claim that the effort isn't necessary, because they have oil reserves for many more decades.

Climate-Change advocates will argue that the proper approach is to phase out all fossil fuels as soon as possible. Here they will run into a fundamental problem, namely that world shortages of fossil fuels means that the worst IPCC cases are no longer credible.

E. Other Impacts

The high-level surfacing of the peak oil problem will negatively impact the stock market, which doesn't like uncertainty. My guess is that an initial market selloff will be followed by a period of reflection, followed by a period of defining new winners, followed by a market recovery. No one likes to see a stock market decline, but it's hard to see how it's unavoidable.

Chapter XII. Conclusions

Discussing energy as an overall category can be confusing and sometimes outright misleading. Energy discussions should be broken down into LIQUID ENERGY (primarily oil), SOLID ENERGY (primarily coal), GASEOUS ENERGY (primarily natural gas), and SOLAR ENERGY. **Those calling for solar technologies to replace oil don't know what they're talking about.**

Energy and related by-products in your day are ubiquitous, well beyond what most people recognize. Breaking a person's everyday life down into electricity, which comes from a number of sources, and oil/natural gas, yields a useful understanding and appreciation of energy-related technologies.

The major products made from oil can be broken down into fuels, asphalt, and miscellaneous. **Miscellaneous oil products include things that are essential to our very existence,** including pharmaceuticals, as well as plastics, paint, bug killers, ammonia, vitamin capsules, trash bags, etc. Oil/natural gas products are everywhere in our lives.

The history of mankind is in many ways the history of energy, which was essential to the development of modern society. For a long time energy choices were primarily a function of the marketplace. Recently, governments have interceded into our energy system decisions in the name of climate change mitigation.

Competent scientists have warned us that climate change is a serious problem, threatening the earth on a multidecadal time scale. Mitigation of climate change is a major dilemma for us all. The problem is not one that can be impacted by changes made on a local level, which are tiny on the scale of the world-level problem. Nevertheless, local governments are making or attempting to facilitate climate change mitigation, hurting local residents with higher energy bills while potentially driving companies away to other jurisdictions or other countries, hurting local employment and sometimes even national security.

Many proponents of taking these tiny actions are arguing that countries quickly stop the production and use of all fossil fuels, which, when burned, emit carbon dioxide, impacting the climate. Attempts are being made to do this, primarily in the U.S. and Europe, while **other important emitter countries have "dragged their feet" on taking appropriate actions – effectively a worldwide stalemate.**

In the midst of what seems like endless squabbling within and between countries, there are human-made, controllable mitigation options available at relatively modest cost, called geoengineering techniques. By way of background, it's well established by volcanos that distributing tiny reflecting particles into the atmosphere/stratosphere cools the planet.

There are other related options lumped into the category of geoengineering. Far and away the best hope is a high priority, no-nonsense program to develop and test these technologies to establish the viability of the most effective of them.

Geoengineering technologies provide a basis for speeding climate-change mitigation.

Moving on to world oil supply. World oil reserves are not known for certain, but recent world efforts to find large, new oil reserves have slowed markedly. One reason is the fact that the world has exhausted the finding of large new oil fields. Another is because of lower investments in the oil industry over recent years, spurred by climate change concerns. **The result is that the world now has the twin worries of a looming investment-related oil production shortage, followed by a long-lasting geologically-based oil shortage, which will be an unspeakable tragedy for humankind.**

The advent of these oil shortages will create worldwide chaos, like what the U.S. experienced during the 1973 and 1979 Arab oil embargos but worse. The timing of these events is unknown, because accurately forecasting them is nearly impossible. Nevertheless, it wouldn't be a surprise if an investment-related shortage were to happen within a matter of years.

The first option when a world oil shortage occurs is conservation, cutting down dramatically on lower priority oil product consumption to ensure that a number of essential activities are provided with needed liquid fuels. Think water supply and food production.

Oil product production technologies using coal and natural gas have been tested in the past and are known to work. However, to build these synthetic fuel facilities is a multiple-year task, even with crash program priority.

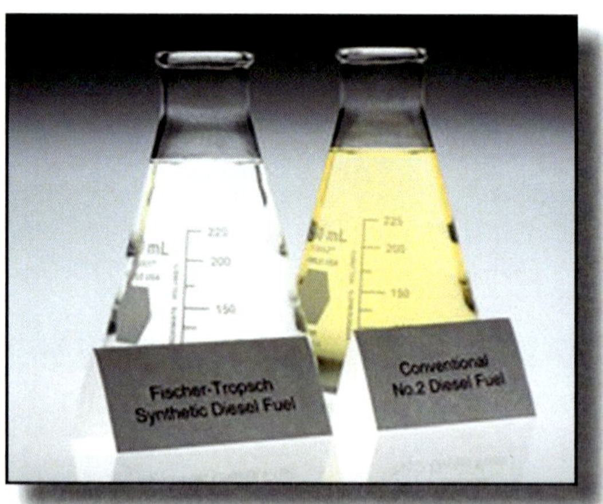

Side-by-side comparison of synthetic fuel and conventional fuel. The synthetic fuel is extremely clear because of its high purity.

Appendix I. Thoughts from Dr. James R. Schlesinger

This is the partially redacted foreword that Jim Schlesinger wrote for our 2010 book, "The Impending World Energy Mess." The primary focus of that book was world oil production peaking. Jim was a gifted intellect, as well as an outstanding public servant. No one else has ever been Secretary of Defense, Director of the Central Intelligence Agency, Chairman of the Atomic Energy Commission, and Secretary of Energy. His message remains insightful and worthy of being my first Appendix.

The ancient Greeks tell us that a Cassandra, despite the truth of her predictions, is doomed not to be believed. Over the years such has been the fate of those forecasting problems for our energy supply, for such forecasts provide discomfort - and political problems from which we prefer to avert our gaze.

Though we readily speak of oil as a finite resource, our tendency has been to pass on to other subjects. Too frequently we are left with the thought that the problem of energy supply will magically disappear. For environmentalists the answer lies in conservation and renewables. More conservative observers think (or hope) that the free market will solve our problems. Yet, even in the face of a supply shortage markets will, indeed, clear. However, the result in higher prices likely will imply reduced national output, the obsolescence of capital equipment, a decline of living standards, and severe political difficulties.

What are the hard facts that lie behind the "finiteness", in particular, of oil supply? One of the founders of the 'peak oil' school, Colin Campbell, observed that in order to produce oil one must first discover it. That sounds like a truism, or a bromide, and it is too easily dismissed as a banality. Yet, when one considers the past and the prospects for oil supplies, it provides a profound truth. Much of our oil production comes from wells that were discovered in the 60s and 70s of the last century. In recent decades we have produced far more oil than we have discovered. In fact, for every three barrels of oil that we have produced, we have found roughly one barrel of newly discovered oil.

The challenge may be simply, if dramatically put. The decline rate from presently producing fields is roughly four million barrels per day. Given the need to replace that production and to increase producing capacity to cover the growth in expected demand, would require the equivalent in the next quarter century of five Saudi Arabias. No one really expects to develop five Saudi Arabias. To this date, we have only found one.

Pessimists believe that we will reach a plateau in oil production in this decade. Optimists believe that we have several decades more before we hit that plateau. The industry generally does not subscribe to the peak oil theory. For the industry the problem is political, the so-called "above ground" problem, rather than geological.

The industry generally contends that there are ample resources, but the nations in which the resources are to be found will not allow efficient exploitation. Nonetheless, as one former CEO has told me, "of course I know there will be a peak, I just don't know when it will come".

The authors provide a detailed analysis of the many proposed alternatives – wind, solar, biophysical, etc. – that might provide physical amelioration for shortages of liquid fuels. It should be noted that the energy problem is basically a problem of liquid fuels. In the early 70s all too frequently the energy problem is treated as if it could be solved by increasing electric power production. Indeed power production could, in principle, be substantially increased, but it would not solve the problem of providing the liquid fuels on which transportation worldwide is dependent.

Some readers will be surprised, even distressed, by the skeptical treatment of global warming. Nonetheless, between the embarrassments suffered by the leakage of e-mails from the Climate Research Unit at East Anglia University, and that the climate models have consistently provided inaccurate forecasts of temperature changes – one should at a minimum - read the skepticism expressed with an open mind. Perhaps even more importantly, one should find it puzzling that it remains the prevailing fashion to devote so much attention to what remains a hypothesis regarding the impact of the release of greenhouse gases on temperature at century's end about which we can do little, given the fuel choices in the developing world, notably China.

Why is it puzzling? The Energy Information Administration projects the increase in the next quarter century of carbon dioxide to be 50%. More importantly is that the projected rise in temperature thus remains quite theoretical and, in the year 2100, many decades away. At the same time, astonishing little attention is paid to the inability to increase oil supply, as demand rises within the next decade or two. Perhaps the reason we devote attention to the former is that the latter is a near term certainty which we prefer not to acknowledge.

Appendix II. Thinking About Our Energy System

A. Introduction

It's essential to think about our energy system in terms of fuels, specifically LIQUID ENERGY, SOLID ENERGY GASEOUS ENERGY, and SOLAR ENERGY. Another useful set of categories is applications, which are TRANSPORTATION, ELECTRIC POWER, INDUSTRIAL, and DOMESTIC.

Clearly (you know this), transportation is primarily powered by liquid fuels, while electric power, industrial and domestic applications are multi-powered.

B. Forecasting

People forecast how energy systems might evolve based on how they envision economies developing, which is increasingly more difficult, the further out in time one goes. The IPCC scenario approach makes sense in that it utilizes differing imagined futures in an effort to bracket reasonable possibilities. However, as mentioned, no significant organization forecasts the advent of wars. If there are small wars, then the large-scale effects would likely be small. If large, then effects could be enormous, requiring decades for recovery, as with World Wars I and II.

No major organization that I know of envisioned the recent Covid epidemic. Yes, plagues can still happen. The overall effects of the Covid epidemic are only dimly understood at this writing. A number of related problems are evident, such as large numbers of deaths, overburdened hospitals, a wide array of supply chain shortages, loss of jobs, shutdowns of schools, increases in crime, etc. The full extent of the damage will only be known after normalcy is reestablished and related analysis is performed.

In the context of energy, -- fossil fuels -- oil, coal, and natural gas – are finite resources; a century of experience tells us clearly that deposits of these fuels are always finite. The more information we have about each fossil fuel, the better our forecasts will be. More on this in our discussion of oil.

C. Short Term Thinking

Most people are not long-term thinkers. The collective "we" tend to worry about today, tomorrow, next week, and next month. Few think years ahead, let along decades. It's human nature. Associated with that aspect of us, many/most of us "glass over" when we hear about things that might happen in the longer term.

Politicians tend to be shorter term thinkers in part because their constituents want them to act on "now" problems. At the time of this writing, demands include: Do something to bring down inflation now; Do something to ensure that Russia invading Ukraine won't draw us into a larger war, now: Do something to get our kids back into school, now; Do something to reduce the crime rate, now; etc. In addition, each political party has priorities that politicians are pressured to help develop and implement. This is not new. Reading U.S. history, we often come across related issues and pressures.

Appendix III. The World Oil Enterprise

A. Introduction

Before we delve into the physical problems associated with future world oil supplies, it's worthwhile to review the structure and realities of the world oil enterprise. A myriad of players are active on an everyday basis. Upstream exploration and production companies deliver oil to intermediate and downstream organizations, including brokers, pipelines, refiners, retailers, transportation companies, and consumers. Buyers and sellers from a number of entities bid for oil every day, establishing price points that bring more buyers into the market or discourage them. Each day brings buy-sell decisions by many organizations. Tongue–in-cheek, one could say it's a well-oiled machine.

Problems can develop when there are disparities between buyers and sellers. While over 110 countries produce oil, the top 10 countries produce over 60 percent of world oil. Add in the next five countries and the top 15 oil-producing countries provide 75 percent of the total. Still others are importers, so the top 15 countries basically control a product that is being consumed by over 200 countries. From a security and price standpoint, this is not an inherently stable situation.

A pumpjack "pulling" oil out of an oil well that is well into production decline.

B. OPEC

While many economists were taught that a cartel was inherently unstable and doomed to failure, the Organization of Petroleum Exporting Countries (OPEC) is an example that has proven otherwise.

OPEC was established in 1961 by five member countries. The organization has expanded since, and two countries have withdrawn their membership because their petroleum imports exceeded their exports. Currently, there are 15 OPEC members. A list of the countries follows:

In Africa:

- Algeria
- Angola
- Congo
- Equatorial Guinea
- Gabon
- Libya
- Nigeria

In the Middle East and Asia:

- Iran
- Iraq
- Kuwait
- Qatar
- Saudi Arabia
- United Arab Emirates

In South America:

- Ecuador
- Venezuela

OPEC's primary goal is to protect the individual and collective interests of its members. According to the OPEC statues, it's tasked to do the following:

- Provide a stable international oil price by decreasing harmful and unnecessary fluctuations,
- Provide for the interests of the producing nations by securing a steady income,
- Provide a steady supply of petroleum to consuming countries, and
- Provide a fair return to those invested in the petroleum industry.

The primary mechanism OPEC uses to achieve its goals is to set production quotas for each member country. By adjusting the amount of oil available to the market, OPEC can in-principle significantly influence oil prices and the profits for its member-countries. The idea is simple but in practice, members have not always toed the line. The temptation for a little extra income has often been too tempting.

C. The International Oil Companies (IOCs)

From its modest beginning in the 1800s, the petroleum industry grew on a foundation of high-risk-taking investors operating in a capitalist environment. John D. Rockefeller overplayed his hand, and the industry became too concentrated, which led to the breakup of the monopolist Standard Oil Trust in 1911. The result was a number of very large super-major oil companies.

From the first half of the 20th century, most of the world's oil production and reserves were controlled by the IOCs, which were vertically-integrated, publicly held companies that explored for, produced, and marketed oil products throughout the world. For a long time, the largest were known as the "Seven Sisters" -- Standard Oil of New Jersey, Royal Dutch Shell, Anglo-Persian Oil Company, Standard Oil of New York, Standard Oil of California, Gulf Oil, and Texaco. These companies were the dominant force in world oil for a long period of time.

Being well organized and well capitalized in what was an inherently very expensive business, the Seven Sisters, during their heyday, were able to pretty much have their way—worldwide. They had access to most of the world's oil reserves and with their large, dominant position in technology, exploration, production, refining, distribution, and marketing, they benefited from increasing worldwide oil demand, which provided large and growing profits for their shareholders. However, their dominance began to fray in the 1960s when a number of Arab countries began exercising increasing control over their in-country oil production through OPEC.

The situation changed dramatically in the 1970s. The IOCs found their opportunities eroded, and they subsequently began to merge and acquire each other in response to the changing marketplace:

- Standard Oil of New Jersey (Esso) merged with Mobil Oil to become today's ExxonMobil.
- Royal Dutch Shell stayed pretty much the same with minor acquisitions.
- Anglo-Persian Oil Company evolved into BP, which subsequently purchased Amoco and ARCO.
- Standard Oil of California became Chevron, which subsequently bought Texaco.
- Gulf Oil was purchased by Chevron with some parts going to BP.

The international oil company influence over the oil business continued to decline as more and more oil producing nations decided to take over and manage their own oil operations through their own nationalized oil companies.

D. The National Oil Companies

Today, many major oil producing countries have nationalized their oil interests, and in the process, "disinvited" the international oil companies from their previous positions and contracts. What used to be called "Big Oil" can be considered "Baby Oil" on a production and reserves ownership basis. Yes, the IOCs are very large economic enterprises, but they've been marginalized by nationalization.

National oil companies typically operate as governmental agencies. They're operated to further the interests of their host countries, without the urgency that's characteristic of the international oil companies, because that's not always in a country's self-interest. Thus, if slower oil development results in larger national incomes because of higher oil prices, and if slower development stretches out their oil reserves, so much the better.

In various cases, national oil company management is highly sophisticated, technologically advanced, and very efficient. The Saudi national oil company Aramco fits that model very well. At the other end of the spectrum is PDVSA, the Venezuelan national oil company, which is choked, starved, and grossly mismanaged by a dictatorial government.

The phase-over from the dominance of the international oil companies to the current dominance of the national oil companies occurred so slowly that most people and governments did not become alarmed. After all, the oil kept flowing and oil prices were generally reasonable until the early part of this century.

In 1970 the International Oil Companies had access to roughly 85% of world oil. Today, the surviving companies have full access to roughly 8% along with some access to another 12%. The shift in control is stark. For those who believe that investor-owned oil companies provide the most reliable, secure sources of oil, the current situation is a source of unease.

Finally, consider ExxonMobil. In the past when oil prices escalated and people felt pain, one of the primary scapegoats was ExxonMobil. It was and is very big and has made huge profits. Because it markets in so many locations, its gasoline prices were and are publicly displayed for people to see on an everyday basis.

When world oil production decline occurs and people again want a scapegoat, be mindful that the primary oil scapegoat in the past – ExxonMobil -- no long qualifies. While huge in scale, ExxonMobil is now a minor player in oil world exploration and production. While some people may reflexively trot out the old scapegoat, those who are informed will realize that they will have to identify a new one.

Appendix IV. More on Oil Production

A. How Various Oil Fields Add Up

Oil output comes from oil fields in varying stages of their lives – some starting up, others on an oil production plateau, and still others in decline. A simple illustration of how production patterns can add up is shown in Figure AIV-1, where hypothetical production from a dwarf, a giant, and a supergiant field are shown. The purpose is to illustrate how production can add up and change over time.

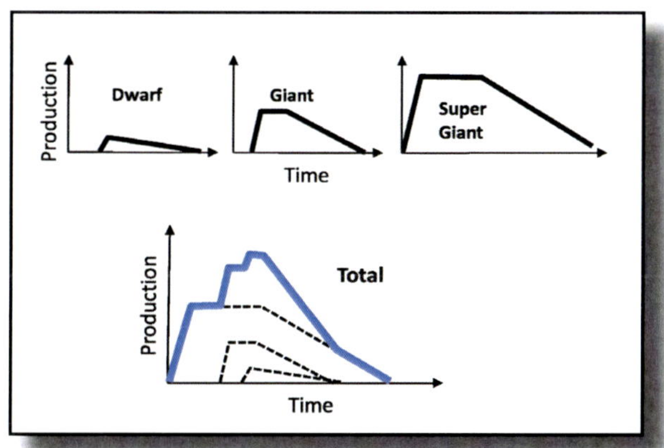

Figure AIV-1. Illustration of how the production from three different sized oil fields adds up. The timing and sizes are arbitrary.

Similar patterns of national oil production ramp-ups and declines have been repeated around the world. In 2005 the Royal Swedish Academy – they bestow the Nobel Prizes in chemistry and physics – did a study of world oil production and noted with alarm that **55 of the world's 65 largest oil producing countries had reached maximum oil production or were in decline**. Since then, Mexican oil production has gone into decline, and Russia may be about to join the declining production club.

With so many countries experiencing oil production decline and the world demanding ever more oil for future growth, the world is in a situation where fewer oil producers are being asked to provide ever more production. At some point that situation will break down.

In 1973 OPEC embargoed oil exports to the U.S., which suffered its first sudden oil shortage. Oil prices increased dramatically as a result. In 1979 the Iranian revolution led to a second major oil shortage, again causing oil prices to skyrocket. Oil prices stayed high until 1986, when they declined precipitously because of a world oil supply glut. So between 1973 and 1986, the U.S. saw much higher oil prices. **If increased oil production was possible, U.S. production should have increased dramatically, according to the high oil price thesis, but it did not.**

Oil exploration and production technologies began a dramatic upswing in capability and effectiveness in the early-mid 1970s, and improvements have continued ever since. If the technology-will-save-us crowd was correct, then the later 1970s and early 1980s should have seen a technology response in U.S. oil production, but it did not.

The reason for oil production declines in certain countries is simple: Oil is a finite resource, which is depleted as it is produced – Production "empties the bucket." When pundents tell us that higher prices and improved technologies can reverse oil production declines, recognize that history tells us otherwise. Indeed, technology advances often allow us to "empty the bucket" more rapidly, rather than recover more oil.

B. What happens to exports whén production declines?

Countries with oil production that exceeds their own needs have exported the excess. But when oil production declines, countries give first priority to their own needs, and exports decline, sometimes rapidly.

Consider the situation in the U.K., which was an oil exporter for more than two decades. As shown in Figure AIV-2, U.K. oil production ramped up dramatically after oil was discovered in the North Sea in the mid 1970s. Production bounced around due to various events and then went into decline around the year 2000. After decline set in, there was no stopping it.

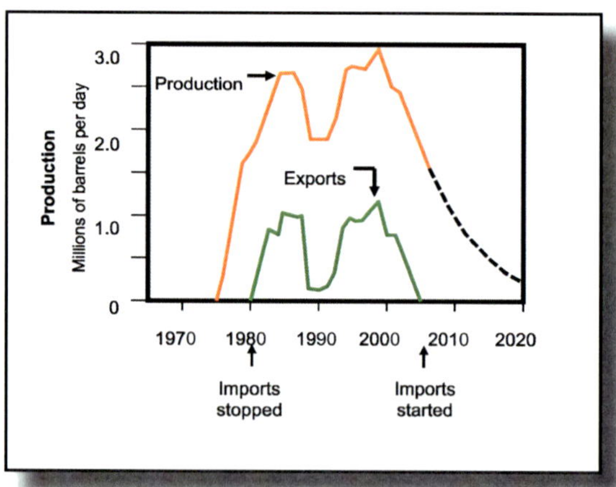

Figure AIV-2. UK oil production and exports. This history shows that when oil production decline sets in, there's no stopping it, in spite of high oil prices and improved technology. Around 2005 the U.K. became an importer of oil again.

Prior to the buildup of its own oil production, the UK was an oil importer. After production rose above its internal needs, it became an exporter. After its production went far enough into decline, its exports stopped, and it became an importer once again. This example illustrates an important sequence of events: **At some point after production in an exporting nation goes into decline, not only do exports stop but imports start and ramp up thereafter.** The loss of an exporter means the addition of an importer.

Various countries have transitioned from being exporters to being importers, and more are sure to follow. When dealing with finite resources, it's inevitable, and **at some point, there'll simply not be enough oil production from fewer and fewer exporters to satisfy the needs of others. That is a simple, irrefutable fact.**

C. The R/P Trap

From the foregoing, it's clear that oil fields behave in ways that are very different from experiences in our everyday lives. That fact makes it difficult for most people to get their minds around the realities of oil production. In the foregoing, we've seen that oil production does not stop abruptly. Rather, it declines over a long period of time, measured in decades. Then why do people talk about years of oil supply?

In various oil producing countries, we have good data on how much oil is produced, and in some situations, we have reasonable estimates of remaining oil reserves. Reserves are measured in barrels, while production is expressed as barrels per unit of time – often years.

Naive analysts sometime divide estimated reserves (R) by annual production (P), which yields a number -- R/P in units of years. They then claim that R/P expresses the years that a country or region will have oil production at a given rate of production. The R/P ratio implies a constant rate of production over a future time period, which is not the way oil production works.

Consider the situation for the U.S. Lower-48 states, shown in Figure AIV-3. The data shows that R/P varied from about 14 down to about 8 and then back up. If the 14 years associated with the 1950 data were adopted as the lifetime of U.S. oil production, production would have ceased by about 1964, which it did not.

People who tell you how many more years of oil production to expect are almost always talking about the ratio R/P. At best, they are uninformed and naïve.

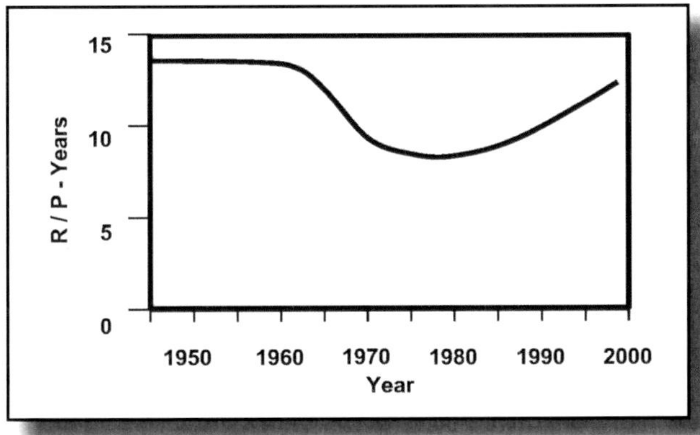

Figure AIV-3. Reserve to production ratio for the U.S. Lower 48 states for the period 1940- 2000.

Appendix V. Oil Production Dynamics

A. Overview

Once new oil fields are found, time, effort, and resources are required to bring them into production, particularly giant and supergiant fields, which require huge investments and long periods of time before meaningful production can commence. Again, it's producing oil fields that provide the oil flows that run our economies. Oil in the ground is important, but oil is only useful when it's produced.

To see how some of these factors interact, consider the history of conventional oil discovery and production in the U.S. Lower 48 states prior to when the shale oil revolution began in 2010 -- Figure AV-1. In this case, peak oil production followed roughly 30 years after the peak in oil discoveries, a time lag often repeated elsewhere in the world.

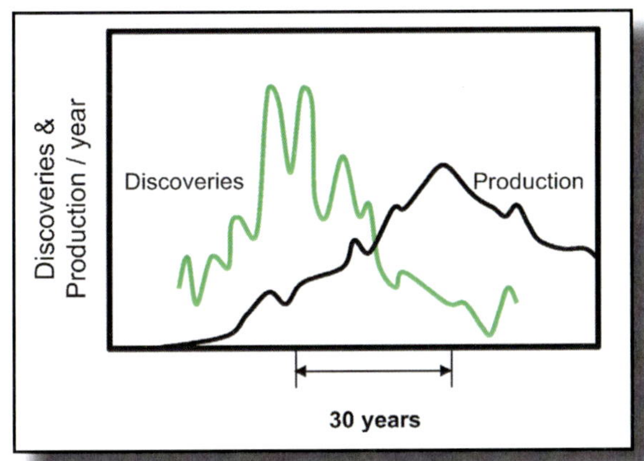

Figure AV-1. The history of discoveries and production in the U.S. Lower 48 states. The peak of oil production occurred roughly 30 years after the peak in discovery.

After the discovery peak, subsequent discovery rates decline, which is not surprising, because there's less and less oil to be found. In almost all cases, bigger oil fields are found early, because they're typically the easiest to find. Smaller fields are found both sooner and later. Keep in mind that finding many more small oil fields does not necessarily result in increased oil production, as illustrated by the U.S. experience.

After oil is discovered, it's almost always immediately developed and brought into production, because of investment realities; oil producers want returns on their investments as soon as possible. As less and less is discovered, less and less will be brought into production, so production will peak at some point after which it will also decline. It's fundamental: **Discovery rates peak and decline, after which production rates peak and decline. The logic is simple.**

B. Giant fields are extremely important.

In the jargon of the industry, the largest oil fields are referred to as Giants or Super Giants – "Giants" for short. Smaller fields are sometimes called Dwarfs. The reserves associated with these three categories are as follows:

- Super Giants – Reserves above five billion barrels.
- Giants – Reserves between half billion and five billion barrels.
- Dwarfs – Reserves up to half a billion barrels.

As already noted, billions of barrels are not as large as they might seem when we consider that oil fields typically produce for decades. Divide a billion barrels of oil reserves by 365 days per year and divide again by a highest production lifetime of say 10 years. The conclusion is that daily oil production rate during the best of times is a much smaller number than the stated reserves number.

Worldwide, it's been estimated that there are somewhere between 50,000 and 70,000 oil fields, the overwhelming majority being dwarfs. Giants are responsible for 50-60% of world oil production. About 25% of total world production comes from just 13 giant fields. **Of the world's 20 largest giants, 16 are past peak production and in decline**. This is not a promising trend!

As earlier noted, the history of giant oil field discovery is not promising, shown in Figure AVI-2. The peak rate of giant oil field discoveries occurred in the 1960s and has been declining ever since, which portends serious trouble in world oil production in the future.

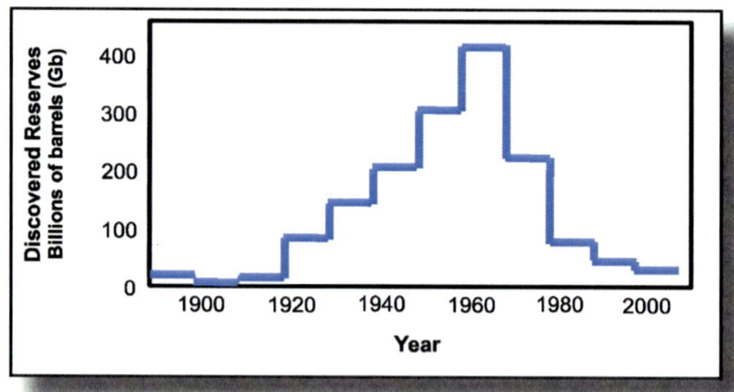

Figure AV-2. Discovery trends for giant oil fields worldwide. Each bar represents discoveries for the decade indicated. The peak discovery rate is seen to clearly have occurred in the 1960s.

C. Recent World Oil Production

Oil production data inherently fluctuate, because of a myriad of forces and events that occur throughout the multitude of oil fields around the world. Included are shutdowns for maintenance, accidents, equipment failures, new fields coming on line, oil fields going into decline, government actions, etc. As a result, to discern a clear breakout trend, we necessarily must wait until oil production moves well outside its previous fluctuation band; otherwise we may simply be witnessing a normal or an unusually large fluctuation, rather than a real trend.

As an example of fluctuations that can occur during a plateau, consider European oil production, shown in Figure AV-3. After a relatively stagnant few years during the late 1980s, production rose to a maximum in the mid 1990s. Thereafter, it fluctuated in a 3% band for about six years. During that period, the data didn't allow us to clearly discern what would happen next. In 2002, production began to break downward, but the trend was not verifiable for several years. It's necessary to wait that long to be sure that the break was not a large negative fluctuation.

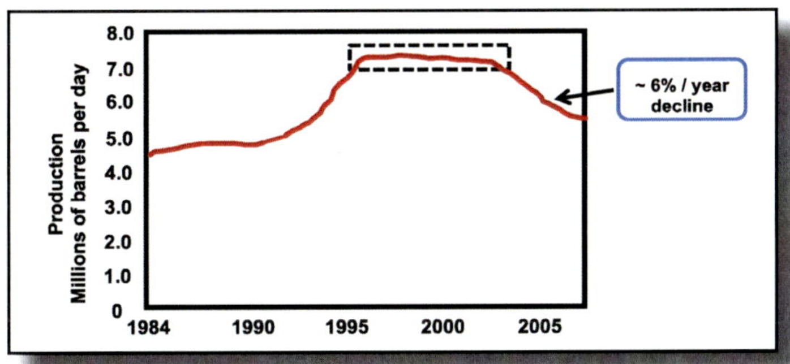

Figure AV-3. European oil production from 1986 - late 2009. Production rose until it hit a fluctuating plateau of roughly 3%, which it maintained for roughly 6 years.

D. Maintaining plateau production is not easy.

Production declines occur every year in various oil fields around the world, so for world production to just remain relatively constant, those declines must be offset by new production. Expansion in world oil supply requires even more new production.

To illustrate the situation, consider three mythical equal sized oil fields that break into a 10% decline at different times, as shown in Figure AV-4. In this fictitious situation, overall production decline rates increase as each field enters decline. The decline rate in this example goes from 0% to 3.3% to 6.6% to 10% over the period considered. On a world scale, the sizes of fields entering decline vary significantly, as does the timing, but the basic pattern is similar to our illustration.

At roughly 100 million barrels per day of world oil production and an arbitrary overall annual decline rate of 5%, the amount of new production needed each year to maintain level production is roughly 5 million barrels per day, which is a very large amount of new production to bring into production each year.

This is an extremely important point. To simply maintain constant world oil production requires adding many millions of barrels per day of new production just to make up for continuing declines in existing oil fields. In other words, we need to run faster just to stay even, let alone increase.

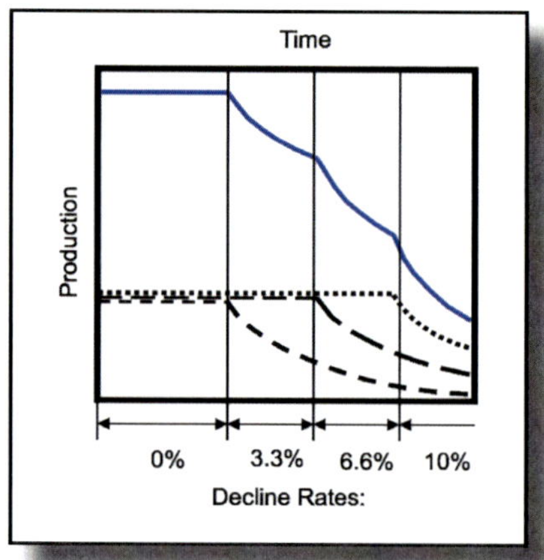

Figure AV-4. Three same-size oil fields that go into declines of 10% per year at different times. As additional fields slide into decline, the overall decline rate increases.

E. The Relationship between World GDP and World Oil Production

The fact that oil is a lifeblood of world economies is evident in the relationship between world GDP growth and world oil production growth over a number of decades, as shown in Figure AV-5. It's clear that as world GDP increased, world oil demand (equal to production) increased in lock step at roughly half the rate.

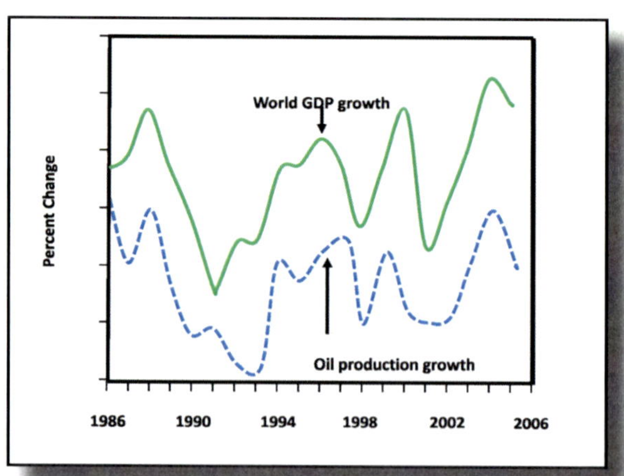

Figure AV-5. Growth in world GDP and world oil production. Over the 20-year period the growth ratio was roughly 2:1, indicating a long history of GDP growth matching oil production growth.

Appendix VI. How Might an Oil Debacle Unfold?

While oil prices have risen dramatically in 2021-2022, they don't clearly reflect the realities of an impending decline of world oil production; rather, they've been driven by the decline in the Covid pandemic and political events in the Russian-Ukraine war.

As already noted, the public and decision-makers have difficulty contemplating a civilization-changing problem that's not obvious, especially when struggling with the recent major U.S. political divide. Besides, decision-makers favor problem clarity, because they then have a clear justification for contemplating serious action.

As a result of these circumstances, it's easy to conclude that preemptive action on the world oil supply problem is unlikely until the reality bursts into the public consciousness in an undeniable manner, at which point it will be too late to avoid extremely serious consequences. How might that happen? Here are a few possible scenarios:

- **Oil prices might escalate to very high levels** as oil shortages develop. When people start hurting badly because of high liquid fuel prices, they will begin to pay serious attention to the fact that something important is happening. This scenario is based on increasing public pain that reaches a critical level at which point the realization of peak oil is suddenly widely recognized.

- **A major political leader outside of the U.S. might announce the problem**. In so doing, that leader would also announce what their government plans to do to mitigate. Such a foreign announcement would almost certainly create a sudden shock and panic, which would spread widely. In this scenario, the U.S. government would be "caught flat footed" and could appear impotent.

- **The U.S. president might announce the problem**. This seems least likely because Administrations are sure to have known about the issue and have avoided or denied it for their various reasons. In this scenario, the President would also have to announce significant mitigation plans, which would not be easy, in part because effective mitigation steps will run counter to the public fixation on global warming, environmental issues, and the push for renewable energy. In this scenario, the U.S. government would seize the initiative and not suffer the embarrassment of being caught unprepared.

Each of these scenarios would create a public shock similar to what occurred in the sudden shortage events of 1973 and 1979. In both situations, the public learned of oil shortages with little forewarning with the following results:

- Rapidly spreading public panic and insecurity.

- Rapid fuel shortages induced by people and organizations rushing to top off their gasoline tanks and beginning to hoard.

- A dramatic escalation in prices for oil products, such as gasoline, diesel fuel, heating oil, and jet fuel.

- Employment reductions by businesses seeking to protect themselves against possible business downturns.

- Large declines in stock markets, as people rush to minimize losses.

- The rapid onset of inflation and recession.

- Increasing interest rates along with bond devaluations.

Clearly, today's economies are different than they were over 40 years ago, but human nature is still much the same, as are the basic tenets of economics. In fact, some of these actions are occurring as a result of the Russia-Ukraine war.

The events of 1973 and 1979 are America's only real-life experiences with sudden oil shortages. The events that ensued back then are not hard to understand; in fact, they're very human. Against this background, when the reality of world oil production decline becomes obvious, there is certain to be sudden, widespread public shock and panic, similar to what happened during the 1970s.

A considerable part of today's U.S. population were not adults during the 1973 or 1979 oil shocks, so they have had no experience with what will be coming. Less experience probably means greater chaos.

Appendix VII. Saving Our Way Out of Impending World Oil Shortages

Increasing the energy efficiency of vehicles, machinery, and equipment fueled by oil products makes excellent sense, as long as the related costs are reasonable and owners can afford the changes. In some cases, retrofits are possible, but in many cases, implementing significant energy efficiency means buying something new.

Another difficulty is that many vehicles, machines, and equipment tend to be large and expensive, unlike computers and electronics, where small unit sizes facilitate rapid changes.

In the following, the U.S. is the basis for discussion; however, most important conclusions are believed to be broadly applicable.

Because transportation is such a large consumer of oil products, its consideration is significant in the context of a world oil shortage. Consider automobiles. One half of recent model year cars will remain on the road over 13 years after purchase. This means that one-half of the vehicles made in 2022 will still be on the road in 2037, and a measurable number will still be on the road a decade later.

For light trucks the median lifetime 14 years; for heavy trucks it's 18 years, and for aircraft it's 22 years.

The largest remaining oil-consuming capital stock exists is in the industrial sector, which is diverse, making it difficult to identify potential efficiency options or potential technology advancements. The largest oil-consuming industries include the chemical, lumber and wood, paper products, and the petroleum industry itself. Industrial use of oil includes process heat, power, feedstock, general heating, and lubrication. **Finally, it's important to note that industry has long been concerned about operating costs, so when more efficient, cost-effective equipment becomes available, they've typically been brought into use on a timely basis. While additional efficiency opportunities undoubtedly exist, it's unlikely that there are huge gains or large energy source changes immediately available.**

In summary, the rapid replacement of less-efficient oil-product consuming vehicles, machinery, and equipment with more efficient replacements would help mitigate the impacts of rising oil shortages and escalating oil prices caused by the decline of world oil production. However, replacement rates normally imply once every 10-20 years and overall costs of trillions of dollars for the U.S. alone. For the rest of the world, the costs would be many times U.S. costs. **In addition, the sudden rush for more efficient equipment will be like people rushing to buy gasoline in 1973 or 1979; production sectors are not built to supply such rush orders.**

In the 2009 U.S. federal "cash-for-clunkers" program accelerated the replacement of over 600,000 autos; it did so at a $3 billion price tag that wasn't sustained because of cost. Such a program is not likely to be viable in the future in a country already facing massive federal deficits that will find itself in a deepening recession caused by oil shortages. It's difficult to conceive of an affordable government-sponsored crash program to accelerate normal replacement schedules to introduce higher energy efficiency and alternate energy technologies into the transportation sector. Significant improvements are inherently time-consuming, because of the size of the fleets of vehicles, equipment, and machinery and the costs of replacements, especially during recessionary times. Another complicating factor is that in times of recession, most consumers will not be able to afford to replace their vehicles – especially since declining oil production will reduce the trade-in values of their existing, less fuel-efficient vehicles. **Our conclusion is that we cannot save our way out of impending oil shortages**.

Appendix VIII. Other Transportation Fuels

A. Introduction

The greatest concerns related to the decline in world oil production are the impacts on transportation, which depends almost entirely on liquid fuels. Possible technologies ready for widespread implementation to help mitigate growing world oil shortages and associated very high oil prices have been described. Other conceivable physical options include natural gas, biomass, and hydrogen.

B. Natural Gas

1. Introduction

Natural gas is used for residential, commercial, and industrial building heating, cooking and drying, electric power generation, feedstock for the production of a number of chemicals, and fueling a few transportation vehicles. Natural gas burns relatively cleanly and has been relatively inexpensive in many parts of the world, at least before the Russian-Ukraine war.

Natural gas is typically transported to end-users by pipelines over short distances and tanker ships over long distances. Natural gas use has expanded in recent decades, particularly for electric power generation, because of its relatively low emissions, and the fact that natural gas-powered electric power plants can be built rapidly at relatively low cost.

The outlook for expanded use of natural gas is promising. It's fueling more of the transportation system is worthy of serious consideration. In the following, some background, reasons for the supply optimism, and considerations related to its use as a transportation fuel are discussed.

2. What It Is

Natural gas is a fossil fuel, primarily methane, a light hydrocarbon composed of one carbon atom surrounded by four hydrogen atoms. Because methane has so many hydrogen atoms per carbon atom, burning natural gas emits less carbon dioxide per unit of energy than coal, which has fewer hydrogen atoms per carbon atom.

When natural gas is produced, it's almost always accompanied by Natural Gas Liquids (NGLs), the light hydrocarbons ethane, propane, and butane, which are liquid under normal conditions. These liquids are typically grouped with oil when considering liquid fuels.

3. Where It Comes From

Natural gas is found in underground reservoirs, either alone or with oil. Natural gas reservoirs are distributed in various locations around the world, often in regions where oil is found. Like oil, natural gas is extracted by drilling wells, which provide pathways for the gas to flow to facilities on the surface that separate out natural gas liquids and impurities, after which the clean gas flows to end users.

Large quantities of natural gas are found in places that are remote from markets. For this reason that gas is called "stranded," because it cannot be economically pipelined to markets. When stranded natural gas is developed, it's typically pipelined to shipping terminals, where it's liquefied by cooling to extremely low temperatures, which facilitates its transport via special tankers. At receiving terminals, the liquefied natural gas is off-loaded and warmed to its gaseous state, so it can be fed into pipelines for transport to end-users.

Liquified Natural Gas (LNG) tankers are used to transport the fuel after it has been cooled to the liquid state.

Like oil, the world's natural gas resource is finite, so it will be exhausted at some future date. However, the remaining resource appears significant. Natural gas is also found in coal seams and in so-called gas shales. Methane in coal mines is a safety hazard, so special care is taken in its handling. In deep coal beds, wells drilled into the coal can extract significant quantities of natural gas, even though the coal is not mined.

As discussed, shale gas is natural gas trapped in shale rock. In recent years it's begun to be widely exploited as a relatively new source of so-called unconventional natural gas, creating optimism regarding future natural gas supply.

Two other sources of natural gas are worthy of mention. Biogas is methane created in landfills, swamps, marshes, sewage sludge, and manure, as a result of the decay of organic matter. Nature has been creating methane in this way for millennia. Biogas normally finds its way into the atmosphere, where it acts as a greenhouse gas.

In landfills, methane can be trapped by overburden and can be extracted for beneficial use via relatively inexpensive, shallow wells. This method of methane capture is practiced in many places and transforms a nuisance into a useful source of small-scale supplemental energy.

Finally, it's worth noting that huge amounts of natural gas occur in the form of gas hydrates, which are ice-like solids in which methane is trapped in a microscopic cage of water molecules. Practical production of gas from hydrates is not now practical, even after decades of research.

4. Natural Gas Use in the U.S.

In residences, natural gas accounts for about 50% of space and water heating. In industry, natural gas is also used for space and water heating, as well as for incineration, metals processing, drying, waste treatment, glass melting, food processing, fueling boilers, and powering large cooling systems. It's also used as a feedstock for the manufacturing of a number of chemicals, fertilizers, and other products. Processing natural gas with steam can produce hydrogen, which is in turn used in oil and chemical refineries.

In electric power generation, natural gas has become increasingly popular, because gas turbine power stations can be relatively easily permitted and rapidly constructed. They can produce relatively low-cost electric power, when natural gas prices are low. Electric power from natural gas generating plants can be easily and rapidly cycled up and down at reasonable costs, which is of particular value to compliment the down cycles in wind and solar cell electric power generation.

5. Natural Gas from Shale

Shale gas is natural gas trapped in shale rock, which is fine grained and not very porous. Natural gas cannot easily move between the tiny pores in shale rock, because of the low connectedness of the rock pores, which is called permeability. Shale gas exists in many countries and regions, including Australia, Canada, China, Europe, India, and the United States.

It has long been known that there are very large amounts of natural gas in shale formations, and shale gas has been produced in small quantities for over a century. Because the large-scale extraction of natural gas from shales has been inherently more expensive than extraction from conventional natural gas reservoirs, shale gas has been of little interest in years past.

The techniques for producing shale gas include vertical and horizontal drilling and fracturing. In simple terms, the sequence involves drilling vertically down into shale rock – often many thousands of feet underground -- and then turning the drill bit horizontally and drilling very long distances out into the shale rock – sometimes beyond a mile. After drilling such a long, L-shaped well, fracturing is used to crack the shale rock at frequent intervals along the horizontal portion of the well. The resulting fractures are held open with propping materials, providing channels that natural gas can flow to. The fractures are frequent enough that much of the gas has a relatively short distance to travel through the shale to reach an open channel, which connects to the horizontal well, through which it can flow up to the surface, where it's processed for beneficial use. Horizontal drilling and fracturing for shale gas production represent advanced developments of technologies that have been in use in the oil and gas industry for decades.

Because it's so difficult for natural gas to flow through low permeability shale rock, shale gas production from newly fractured wells is very high during the first year of operation and declines dramatically thereafter. This is the result of gas close to fractures being able to exit quickly, while gas further away from fractures takes longer to reach fractures. Production from shale gas wells can decline 50 percent or more in the first year, which is dramatically faster than the flows of typical conventional gas wells, which flow at high levels for long periods of time. Shale gas wells are kept in operation for many years, but the greatest production occurs in the first year. The situation is a little like a person taking a deep breath and exhaling hard – There's a sudden burst of air followed by less and less air as lungs collapse.

In practical terms, this pattern of shale gas production means that maintaining high overall levels of shale gas production requires the continuous drilling of new shale gas wells. In principle, continuous drilling is possible. However, since natural gas prices are typically volatile, shale gas operators may not make enough money during periods of low prices to maintain continuous drilling. The resulting volatility can lead to aggravated boom and bust cycles, which would not be in natural gas investors' interest.

The issues associated with the massive development of shale gas are cost, water use and disposal, fugitive emissions, and the need for continuous, uninterrupted development.

6. Methane Hydrates

Methane hydrates are ice-like solids in which methane is trapped in a microscopic cage of water molecules. These hydrates exist in many locations worldwide in cold regions, such as the arctic, and underwater, where temperatures are higher but where high pressures maintain the hydrates. Recall that methane is the primary component of natural gas.

The hydrate resource is enormous but not estimated with great accuracy. However, "enormous" is enough to capture people's interest and to justify research on how methane hydrates might beneficially be produced.

Without cap rocks or other containment, methane released from hydrates will go where it wants. Since it's a light gas, methane tends to move vertically away from its former place of residence, eventually escaping into the atmosphere, where it is a potent greenhouse gas. That reality doesn't facilitate its easy capture; it could create a fire hazard, creating carbon dioxide enhancing the greenhouse effect. The absence of a cap rock or other containment structure thus represents a major impediment to the practical capture of methane hydrates. Nevertheless, because the methane hydrate resource is so large, related research is certainly justified.

7. Natural Gas For Transportation – The Chicken and Egg Problem

Readers may have noticed natural gas-powered buses and taxicabs in various cities, where its use is proudly advertised for all to see. Besides being economical, natural gas-powered buses avoid that unsavory smell often associated with older diesel fueled buses.

Natural gas buses have established routes, and refueling is facilitated by the fact that the buses typically return to their garages each night, where they can be refueled. Hub-based operations are necessary, because natural gas fueling stations are rare. Therein lies one problem with moving quickly to deploy natural gas vehicles, other than buses, taxis, and other special situations -- namely the lack of an available refueling infrastructure, "Chicken and egg."

There is potential for natural gas in fleet markets such as delivery trucks and vans, government vehicles, school buses, etc. Many of these vehicles operate within relatively small geographical regions, have defined daily routes, and return to the same central location where they can be refueled.

One intriguing option is to convert school buses to natural gas fueling, which is feasible because the buses typically return to a central location, where they can be refueled. With the advent of a national liquid fuel shortage, these buses could be pressed into use for general public transportation, which would provide a ready option for the mitigation of the decline of world oil production. A significant problem is that most school districts can't afford the conversion of their school buses, so state and federal governments would have to become involved.

Before a country undertakes a large national commitment to natural gas vehicles, it would be prudent to establish that there is an assured long-term supply of relatively low-cost natural gas available on which to base natural gas vehicle development. However, it must be recognized that there are other options for vehicle fueling, so it's essential to objectively consider all options to determine which one or ones are likely to be the best choices.

C. Biofuels

1. Introduction

Liquid fuels that can substitute for oil-derived fuels can be produced from biomass. Biomass is defined as plant material such as trees, grasses, corn, wheat, soybeans, palm, sugar cane, algae, etc. The earliest biomass-based energy was burning of wood for heat and cooking. The current interest in biomass energy is associated with the fact that it's renewable, which means new crops can be produced on a regular basis. In the past, the thought was that biomass energy had very positive environmental impacts, but recent studies have raised questions.

In the context of this book, the interest is in biomass-to-liquid (BTL) fuel options. In the U.S., there is a large, on-going effort to grow corn and transform it to ethanol, which can be used as a gasoline substitute. In Brazil, sugar cane is grown on a large scale for the production of ethanol. Other crops are also being grown for conversion to fuels in various countries.

Before discussing some of the specific biomass-to-liquid fuel options, we discuss a few fundamentals associated with biomass plantations – crop areas that are specifically dedicated to the growth, harvesting, and processing of biomass for the production of liquid fuels.

2. Some Biomass Plantation Fundamentals

Growing and harvesting large quantities of most land-based biomass requires large land areas. Harvesting biomass requires tractors and trucks that consume liquid fuels, which means **that biomass-to-liquids is inherently a process that consumes liquid fuels in order to produce liquid fuels.** This translates to an inherent limit on the size of land-based biomass plantations – a limit that's determined by the fuel consumption required to harvest and transport biomass that is a long distance from facilities that can convert biomass to liquid fuels.

Consider the factious case of converting the total land area of a dozen U.S. Midwestern states to biomass-to-liquids production. That task would require wiping out all major cities, factory areas, and large parks, and devoting nearly all of the land for biomass-to-liquids production. This is clearly a fiction, but the purpose is to calculate some illustrative numbers to provide some overall perspective.

For this calculation, consider the total land areas of the following states: Michigan, Minnesota, Kansas, Nebraska, Oklahoma, Missouri, Wisconsin, Illinois, Iowa, Arkansas, Ohio, and Indiana. The sum total land area of these states is near 800,000 square miles, which includes lakes, rivers, hills, mountains, and unusable areas, which we ignore for simplicity since it's a fictitious case anyway.

Assuming an optimistic, steady liquid fuel production rate of 8,000 barrels per day from a 50-mile radius biomass plantation yields an approximate yield of a barrel per day per square mile over the course of a year, which is a rough number. The important point is that typical liquid yields are not as low as 0.1 barrel per day or as high as 10 barrels per day.

The bottom line is that the total production from biomass-to-liquids plantations covering all 12 states would be roughly 800,000 barrels per day, which is roughly 4% of recent U.S. liquid fuels demand. **On the basis of this fictitious but optimistic illustration, it's not remotely reasonable to believe that land-based biomass- to-liquids processes can provide significant mitigation of the decline in U.S. oil production.**

3. Corn-Ethanol: A Demonstrated Loser

Ethanol can be mixed with gasoline to form a fuel mixture that works in existing vehicles, built to operate on gasoline alone. In the U.S. ethanol is blended with gasoline at the 10% level and widely marketed. To use ethanol in higher concentrations in U.S., automobiles and light-duty vehicles would require vehicle redesign, because current vehicle materials would lead to accelerated engine failures.

Ethanol is the alcohol found in alcoholic beverages. It can be easily produced from sugar cane and corn by fermentation and distillation, which is the approach responsible for most of the ethanol produced in the world today. Brazil is blessed with growing conditions in which sugar cane flourishes, and that country has taken advantage of that fact to produce ethanol on a very large scale. The situation in Brazil is economically and environmentally reasonable. In addition, sugarcane-ethanol is energetically attractive, in stark contrast to corn-ethanol in the U.S. Finally, vehicles in Brazil are built to run on high concentrations of ethanol.

The U.S. produces massive amounts of corn-ethanol, but the Energy Return on Energy Investment (EROEI) for the process is poor -- some claim that it's negative, meaning that more energy is required to produce corn-ethanol than is in the ethanol product. Massive subsidies and requirements for corn ethanol in the U.S. are the result of politics associated with the farm states, not sound energy policy.

The interest in corn-ethanol was rooted in the fact that it can be grown, which makes it renewable, which is a high value to environmentalists, politicians, and the public. Farmers are particularly fond of corn-ethanol because it represents a major outlet for corn production, enhanced by subsidies and mandates.

Corn-ethanol requires the use of very large areas of fertile, arable land, which would otherwise be used to grow food. U.S. mandates for corn-ethanol caused the price of food to escalate, an unwanted outcome that could have been anticipated, had there been careful analysis before the corn-ethanol mandates were imposed. At the beginning of the corn-ethanol craze, few analysts suspected that corn-ethanol would also have significant environmental problems. After all, it was renewable, which implies goodness. **Today, we know that corn ethanol is environmentally marginal, which simply adds to its negatives.**

The corn-ethanol craze in the U.S. is the result of politics moving ahead of objective analysis. Reversing the situation will be difficult because of farm state interests, and maybe the fact that the first U.S. presidential election caucus is held in farm-state Iowa every four years. No wonder so many in the real world of business fear political intrusion into energy markets.

4. Cellulosic Ethanol

Cellulose is carbohydrate plant material, which is the main constituent of cell walls in wood, grasses, cotton, and other plants. Because plants are renewable, cellulose is renewable. If ethanol or hydrocarbon fuels could be made from cellulose in an economic, energetically, and environmentally reasonable manner, cellulosic ethanol could make a useful contribution to future liquid fuel supplies.

Using a variety of feedstocks, cellulosic ethanol and other fuels could be produced in many regions of the world. Sources of cellulose are myriad and include agricultural wastes, industrial wastes from plant materials, and energy crops, like trees and various grasses, grown specifically for fuel production. Note that for a number of these materials, there are existing uses, including fertilization of existing soils, drywall, wallboard, and animal feed. Removing those cellulosic materials could cause upsets in markets and the environment. Accordingly, related changes in use would need to be carefully evaluated. **Rarely do we get something for nothing.**

Growth of cellulose does not require fertile land, so it need not compete with the growth of agricultural foods, which is an important plus. **The problem is that liquid fuels made from cellulose are a long way from being economic, because of the need for complex, difficult, and expensive conversion processes.**

One of the important questions concerning the potential viability of cellulosic-based liquid fuels is whether the related returns on energy and liquid fuels are sufficiently positive for this approach to be viable. Some studies have indicated that cellulosic ethanol is promising, but many studies were not peer-reviewed for objectivity and accuracy, and some studies were conducted by vested interests, such as renewable and agricultural fuels organizations, environmental organizations, and units of government with potential conflicts of interest.

It is still very much an open question as to whether cellulosic-based liquid fuels can have large enough net benefits to be of practical value. It is also likely that some countries will have greater potential than others.

5. Algae-To-Liquids

Another biomass-to-liquids concept receiving considerable attention is Algae-To-Liquids (ATL). In one sense, this is an attractive option, because **algae make very efficient use of energy from the sun and various nutrients.** However, ATL has a fundamental liquid fuel return-on-investment problem that must be overcome.

Algae grow in water and are thoroughly soaked when recovered. A great deal of energy – usually liquid fuel energy -- is required to harvest the heavy, water-soaked algae and to remove the water. This presents a fundamental roadblock, which, to my knowledge, has yet to be overcome.

It might be possible to develop algae that secrete hydrocarbon liquids that are easily collected. If so, the dewatering problem could be circumvented. Since there has been a good deal of venture capital investment going into algae-based liquid fuel production, there may be other research options possible.

6. Concluding Remarks

Most biomass production is currently committed to food, feed, lumber, paper pulp, fiber, etc., and biomass makes heavy use of fossil fuels for planting, tilling, harvesting and transportation of material to and from processing.

Compared with current liquid fuels use, the potential impact of successful biomass-to-liquids development would be relatively small in many locations in the world. Increasing the average mileage of internal combustion, passenger vehicles by 3 -5 miles per gallon would dwarf the effects of all possible biofuel production from all sources of biomass available in the U.S.

Nevertheless, because all reasonable sources of liquid fuels will be needed to mitigate the impending decline in world oil production, further research and development of biomass-to-liquid options deserves attention. A number of successful biomass plantations, operating for a number of years would be needed to clearly demonstrate viability. Let's not "bet the farm" until clear winners have been demonstrated and clear losers discarded.

D. Hydrogen

1. Overview

Hydrogen has significant potential to provide energy for various transportation applications. Many envision hydrogen as a long-term alternative to petroleum-based liquid fuels.

Hydrogen is an energy carrier, not a primary fuel such as oil, natural gas, coal, wood, etc. Hydrogen production requires investing energy from a primary fuel to separate it from some other material, typically natural gas or water. Electricity is also an energy carrier in that its creation requires a primary fuel.

Hydrogen can be directly combusted in internal combustion engines, like those now in use, or it can be fed to fuel cells, which directly and efficiently produce electricity. An automobile powered by hydrogen fuel cells is a type of electric vehicle in which the large battery pack is replaced by a hydrogen storage tank and fuel cells. Such a vehicle must be periodically refueled with hydrogen at service stations, only a few of which now exist on an experimental basis, primarily in California.

Fuels cells have no moving parts. They combine hydrogen and the oxygen in the air to very efficiently produce electricity and their emissions are water vapor.

The challenges in going the hydrogen route include 1) the need to generate hydrogen economically; 2) transporting the hydrogen to service stations, where users can safely refuel their vehicles; 3) building and operating hydrogen service stations; 4) improving fuel cells so that they are economical and able to operate effectively in a wide range of vehicle environments; and 5) developing hydrogen storage systems that can efficiently, economically, and safely operate on vehicles.

2. A hydrogen economy will not happen easily

As with natural gas use in vehicles, development of hydrogen vehicles involves a "chicken and egg problem." On the one hand, building a widespread hydrogen distribution system with service stations represents a huge, expensive undertaking. On the other hand, selling large numbers of hydrogen vehicles requires that service stations be in place for easy, convenient refueling. Hydrogen generation and distribution systems must come first, but they will initially have a small number of customers, since initial sales of hydrogen fuel cell vehicles (HFCVs) will be small compared to the fleets of other vehicles on the road. The bottom line is that huge subsidies will be needed to start a HFCV program.

Hydrogen is currently produced on a massive industrial scale for the refining and chemical industries. The cost of hydrogen derived from natural gas is comparable with the recent cost of gasoline. However, fuel cell development has a ways to go before fuel cells can provide the performance needed in vehicles at costs that consumers might be willing to pay.

Safety is an issue to be addressed also. There's a big difference between an excellent hydrogen safety record achieved by well-trained professionals in industry and safety in the hands of everyday consumers. Hydrogen is flammable and explosive, so precautions are required at all stages of hydrogen transport, storage, and handling in consumer settings. Considerable effort and years of experience will be required before systems can be demonstrably considered safe.

While hydrogen fuel cell vehicles are a potentially attractive long-term technology option for displacing gasoline and diesel fuels in transportation, the road ahead in hydrogen will be challenging. Optimism that the technology will succeed is justified, but it won't happen quickly, so hydrogen fuel cell vehicles are not a likely near-term option for the mitigation of the decline of world oil production.

Appendix IX. Administrative Mitigation

NOTE: This Appendix is a redacted reproduction of a similar section of our 2010 book, "The Impending World Energy Mess," by Hirsch, Bezdek and Wendling. Parts of the following may be somewhat dated in that we've learned much from adjusting to the Covid pandemic that applies to how we might manage certain aspects of peak oil mitigation, e.g., remote working, food delivery instead of traveling to the grocery store, etc. I've added "chaos" comments in a few of the places where chaos is almost certain.

A. Introduction

When world oil production goes into decline, there will be annually growing oil shortages and much higher prices for gasoline, diesel fuel, jet fuel, home heating oil, various chemicals, lubricants, etc. People, companies, and governments will attempt to do everything in their power to mitigate the myriad problems that ensue.

Consider two classes of mitigation. The first is Administrative Mitigation, which covers options that can be implemented by individuals, organizations, and governments, voluntarily or via mandates. The second category is Physical Mitigation, which includes construction and deployment of new vehicles, equipment, machinery, and liquid fuel production techniques and production plants.

Administrative options are considered first: Government rationing of oil and oil products, car-pooling, and telecommuting. Second, we discuss physical mitigation.

B. Rationing Options – A Rat's Nest of Complications

1. Introduction

When oil shortages develop and prices escalate, government intervention will happen and be required. Rationing to moderate the effects of the oil shortfall is an important but extremely complex undertaking.

There are four generic oil and oil product allocation options that could be implemented: (i) oil price and allocation controls, (ii) coupon gasoline rationing, (iii) a variable gasoline tax and rebate, and (iv) no oil price controls with partial rebates. The choice of appropriate policies is by no means obvious. Furthermore, their implementation will be difficult and imprecise, and they cannot eliminate negative effects on the larger economy. Our discussion focuses on the U.S. but is applicable to most developed nations.

Many readers will have difficulty following the complexity, intricacies, and unintended consequences of these approaches to the non-market allocation of oil shortfalls. The onset of the decline in world oil production will cause widespread economic hardship and social unrest and will be accompanied by demands by many that governments "do something" to remedy the situation. Rationing is certain to be an option on the table. Government rationing and redistribution of oil products is inherently complex and controversial and raises its own set of problems, costs, and inequities.

As in previous oil shortages (1973 and 1978), unless the government imposes price controls on domestic oil production, oil prices will rise with the world price, and the increases will be passed on to consumers. For the shortfalls likely to occur when world oil production goes into decline, the size of oil price increases is impossible to estimate, but increases of 100 to 300 percent are certainly possible, assuming significant prices at the outset.

Oil shortages will reduce a nation's production of goods and services directly. Industries will be forced to reduce their oil use, resulting in lower output and higher prices for their products. Any approach that allocates a shortfall of oil cannot offset the supply-side costs, although efficient allocation could minimize them. Further, as workers seek to protect their real incomes, higher prices for oil products could be reflected in demands for higher wages, setting off a wage-price spiral that would compound the inflationary effect of higher oil product prices. Outlays for programs indexed to price levels (such as cost-of-living contracts and pensions) would rise automatically.

2. Criteria for Comparison

There are several criteria for comparing oil-rationing approaches; here are four:

1) **Microeconomic (small-scale) effects**. Each approach will affect the prices of goods and services, inventory management of manufactured goods, and inter-industry and inter-regional relationships.

2) **Macroeconomic (large-scale) effects**. Each approach will affect national economic growth, inflation, and other important macroeconomic variables.

3) **Equity**. Some parts of national economies will benefit and some will lose. People in similar situations may not be treated equally. Rationing approaches may be perceived as fair or unfair across different income groups and regions of a country.

4) **Practical problems**. This includes all the costs of implementing each approach. For example, regulations to prevent fraud and abuse may be required with an allocation approach that distributes gasoline coupons or cash. What impacts on each approach will lobbying activities have? Will politically powerful constituencies develop and influence implementation?

3. The Workings of Each Approach

Price and allocation controls. This approach could impose domestic oil price controls, an entitlements program, and regulations similar to those that were in effect in the U.S. from 1973 through January 1981. Price ceilings were imposed on domestic oil production so that prices would not rise with world oil prices.

Since some refiners had access to price-controlled domestic oil while others were forced to buy world-priced imported oil, the government imposed an entitlements system, averaging the prices of domestic oil and imported oil, so all U.S. refiners paid approximately the same price. Thus, even under price controls, the average price of crude oil rose. Controls on the price markups of downstream operators (crude oil resellers, refiners, wholesalers, and retailers) were also required; otherwise, they would be able to raise prices to market-clearing levels and thus negate the effect of price controls.

Allocation regulations were imposed in the U.S. under which suppliers were obligated to sell proportionately reduced volumes to their historical purchasers. Thus, if a wholesaler were able to satisfy only 80 percent of its demand for residual fuel oil, the supplier reduced deliveries to all historical residual fuel oil customers by 20 percent. The government designated priority users, who received a higher portion. For gasoline and diesel fuel, however, the allocation system stopped short of retail purchasers; that is, consumers were not allocated specific quantities. Each gasoline station received reduced supplies of gasoline based on historical purchases, so the demand for gasoline at the controlled price was greater than the quantity supplied, so cars waiting in gasoline lines resulted (Figure AIX-1).

Figure AIX-1. U.S. gasoline lines during the oil crisis of 1973.

Further actions were required when refiners produced less of a given oil product than deemed desirable. This could occur in the case of home heating oil, and the government had to direct refiners, through refinery yield orders, to change their relative yield of products -- for example, to produce less gasoline and more fuel oil.

Gasoline coupon printed but not issued during the 1979 energy crisis

Coupon gasoline rationing. This option would impose coupon rationing for gasoline allocation controls, but it would retain the other controls described above. It would limit price increases for gasoline, as would the first option, but would use ration coupons rather than queues to allocate gasoline. Coupons could be distributed on the basis of registered vehicle ownership, on the basis of historical use, and to priority and hardship users. Since coupons could be sold, market forces would set coupon prices.

Price controls would ensure that the price of gasoline do not rise to market levels. Gasoline price controls do not set price ceilings but instead control profit margins, so price controls on domestic crude oil would be required to keep the average refiner acquisition costs of oil from rising as quickly as world oil prices. Without domestic price controls and margin controls on downstream operations, gasoline prices would rise to market levels, causing ration coupons to decrease in value over time and eliminating the need for rationing. Without price and allocation controls on oil products other than gasoline, refiners could increase revenues by increasing the prices and production of these products. A system of local boards would be established to administer state ration reserves, providing additional allotments to those who would otherwise experience severe hardships. **(Think chaos!)** Ration coupons would be required for the purchase of gasoline and would then be transferred from retailers up the distribution chain to refiners and finally back to the government.

Gasoline tax and rebate. This approach would consist of a system of emergency gasoline taxes and rebates structured to have effects similar to gasoline rationing with a free market in coupons. Price and allocation controls on gasoline would be prohibited, but they would be required on oil and oil products to keep refiners' oil costs from rising to world levels and to prevent downstream margins from rising. The level of the variable tax on gasoline would be set so that refiners could pass through increases in average oil prices. Thus, as world oil prices and gasoline demand changed, the size of the tax would change. Tax revenues would be rebated to registered motor vehicle owners, just as coupons would be distributed under rationing. **(Think chaos!)**

A specific example illustrates the similarities between the two approaches. Under the rationing approach, the government would control the gasoline price at, say, $5 a gallon. A ration check (or electronic transfer) for a set number of ration coupons, each good for one gallon, would be sent to each vehicle registrant, with a limit of three registered vehicles per household. Assume that each car owner would get 10 coupons a week. Persons requiring more than 10 gallons a week could purchase coupons from those needing less than 10 gallons. As a free market for ration coupons develops, the going price might settle at about $6 a coupon under a 20 percent oil shortage. Thus a gallon of gasoline could be purchased for $5 and a coupon worth $6. The market-clearing price of gasoline would be $11 a gallon -- $5 for a gallon of gasoline plus $6 for the coupon.

Under the gasoline tax and rebate approach, the results would be similar. For a shortage of the same size, the market-clearing price would still be $11 a gallon. An emergency gasoline tax of $6 a gallon would add $6 a gallon to the oil distribution chain. The tax and rebate approach would permit the oil distribution

General rebate. Here the prices of all oil products would be allowed to increase. Prices would determine how much of each product would be produced and how the products would be distributed, and domestic oil prices would be allowed to rise. A windfall profits tax would capture a large portion of the higher oil revenue, which would then be rebated. Rather than trying to avoid price increases by using price controls, this option accepts the price increases and attempts to offset the negative economic consequences with rebates. Whereas gasoline consumers would be most directly affected under the preceding two approaches, all consumers would be affected under the general rebate approach, which would presumably distribute the rebates more broadly. The distribution mechanism could take many forms, such as adjustment in federal income tax rates and changes in withholding liabilities, changes in existing transfer payments, and reduction in the federal debt. It is worth noting that "windfall profits" taxes were implemented in the U.S. between 1980 and 1987, and the results were hugely counterproductive, according to a 1990 Congressional Research Service report. "The WPT reduced domestic oil production between 3 and 6 percent, and increased oil imports from between 8 and 16 percent," says the report. "This made the U.S. more dependent upon imported oil." It took many years after the tax was terminated to settle all of the issues that the U.S. windfall profits tax created.

The recessionary effects of a liquid fuels shortage will reduce tax receipts in the non-energy sectors of the economy. Reduced revenues, combined with increased unemployment and other transfer payments, will create budgetary demands that could be financed by the increased income from the windfall profits tax. Some portion of the rebate could also be used to assist firms by reducing corporate income taxes; businesses, particularly oil-intensive ones, would have to raise the prices of their products as oil prices rise. Because consumers would receive income supplements in the form of rebates, most businesses would be able to maintain their market. Some, however, would not. For example, California fruit growers who rely on truckers to transport their produce to East Coast markets would find their goods less competitive with fruits produced closer to the market.

4. Microeconomic Effects

Price and allocation controls. Under this approach, prices would not rise to market levels and queuing would result, because of oil product shortages. However, the true price of gasoline, including the cost of waiting in line, would rise to market-clearing levels, and the "waiting cost" would reduce consumer welfare just as would a gasoline price increase. The difference is that without price controls, a higher dollar price is paid to others in the domestic economy and does not involve a loss in well-being for a nation as a whole. An increase in the effective price of gasoline to consumers produced by queuing is a net loss to society, a loss that could be equivalent to hundreds of billions of dollars in the U.S., for example. Further, to the extent that government-determined allocations diverged from allocation to highest value uses, significant losses in national economic efficiency would also be incurred. An allocation rule based on historical use would not be able to keep up with changing patterns of demand or determine which customers could reduce consumption most efficiently. In addition, government officials would be required to decide the priorities of oil distribution, and interest groups would likely influence these decisions. **(Think chaos!)**

Inventory behavior would be directly affected by the allocation approach selected. To the extent that future profits from storing oil would be limited during a disruption by price controls, rationing, or taxes, less oil might be stored by the private market prior to the disruption. During the shortfall, price controls would delay oil price increases; thus it would be in the inventory holders' interest to hold stocks while the price increases. Without controls, however, the unconstrained price would increase rapidly. After reaching the market levels, prices may stabilize, reducing the incentive to maintain high oil inventory levels.

Coupon gasoline rationing. Coupon rationing would eliminate gasoline lines and allocate gasoline supplies to government specified highest value uses. However, this approach would introduce inefficiencies into the oil market. If the bulk of shortfalls were borne by gasoline, the approach would allocate other, underpriced refined petroleum products to those who might otherwise conserve. Very expensive conservation measures might be forced on gasoline consumers, while relatively inexpensive conservation efforts for other petroleum products would be forgone. For example, commuting by automobile might be made prohibitively expensive, especially for low-income persons, rural households, and those with very long commutes, and industries that depend on automobile traffic would be severely affected. In contrast, if the shortfall were distributed across all oil products, all consumers would find more efficient ways of conserving petroleum products, eliminating the need for gasoline consumers to take extreme measures. A system of price and allocation controls would be imposed, resulting in the inefficiencies described earlier. Finally, no incentives to increase domestic production would be provided by gasoline coupon rationing.

Gasoline tax and rebate. In two respects, this approach is similar to gasoline rationing: (i) the losses incurred if the shortfall were borne primarily by gasoline would be the same, and (ii) the losses from imposition of price controls would be the same. In both cases consumers do not face market-determined prices for gasoline, and incentives to increase domestic oil production are absent. However, because the tax and rebate approach would not require gasoline price controls, it could result in a more efficient allocation of gasoline supplies. The delay between purchases and rebates might impose hardships on many consumers. **(Think chaos!)**

General rebate. The general rebate approach would minimize micro-efficiency losses. By encouraging conservation in the use of all refined products, it would put the available supply of oil to its highest valued uses, and it would avoid socially divisive queues for gasoline.

5. Macroeconomic Effects

Price and allocation controls. Like the other options, this approach would not mitigate the supply-side macroeconomic costs of the oil shortfall. The higher effective price of oil would reduce real GDP and raise the general price level. The costs of queuing up for gasoline would be enormous, though not measured directly. For example, in the U.S. in the second quarter of 1979, real GDP decreased 2.3 percent at an annual rate, and much of this decrease was attributed to the oil shortfall. On the demand side, however, oil price ceilings, if effective, would limit the transfer of funds to oil producers and the resulting oil price a drag on non-oil markets. Ceilings on oil product prices would also limit increases in the consumer price index (CPI).

Coupon gasoline rationing. This option would not alter the supply-side effects of an oil shortfall, but by controlling domestic oil prices, it would tend to limit fiscal drag. In contrast to a gasoline tax and rebate approach, the ration coupons (paper or electronic) provide a second currency that may insulate some sectors of the economy from fiscal drag. In addition, if the coupon price were excluded from the CPI, the inflationary impact could be reduced. The microeconomic efficiency benefits over the price and allocation controls discussed above could result in less reduction in economic activity, since the rationing approach permits free-market trading of coupons and would allocate gasoline more efficiently than would direct allocations. As a result, Gross Domestic Product (GDP) would tend to be higher and the price level lower than would occur under price and allocation controls alone. On the other hand, if the shortfall were borne primarily by gasoline consumers, economic activity would be retarded, relative to the situation with no controls.

Gasoline tax and rebate. This approach cannot alter the supply-side effects of oil shortages. A tax-rebate system that includes price controls could limit oil price drag, of uncertain and uneven rebates, the adverse impacts could be sizable. Further, the CPI would directly reflect the gasoline price increase, and this effect would be significant. For example, under the conservative assumption of a doubling of gasoline prices, the CPI could increase by more than five percent in the first month of the program (an 80 percent annualized rate). The increase in consumer prices would trigger increases in indexed wages and entitlement payments. More rapid wage inflation would increase production costs in the economy, and the inflationary impact would be prolonged by second- and third- round effects on wages and prices.

General rebate. This approach would not mitigate the supply-side costs of the shortfall, but by allowing all supplies and demands to interact at market-clearing prices, decontrol would achieve a greater degree of economy-wide efficiency and a higher GDP than any of the alternative approaches. Even if rebates were distributed immediately, some oil price drag would occur, and funds would flow from non-oil to oil sectors. In addition, rising oil prices would increase the CPI. However, the reduction in economic activity must be balanced against the efficiency gains in terms of both resource allocation and administrative costs of a decontrol system.

6. Equity

Price and allocation controls. Under price and allocation controls with queuing, there is a transfer of income from those who value their time more than the average to those who value it less or have no choice. This transfer of income may be monetized as the size of the queues increase, as persons with a high value of time may pay others with a low value of time to wait in line. Further, it is not clear that historical allocation is "fair," because regions of the country that are growing more rapidly than others would probably feel they were being treated unfairly. Because of the shift in consumption and driving patterns over time, allocation on a historical basis would become more and more unfair.

Coupon gasoline rationing. The fairness of this proposal obviously depends on the distribution of the coupons. If fairness means reestablishing an individual's purchasing power prior to the disruption, a rationing approach should distribute the coupons according to the amount of gasoline consumed prior to the disruption. Alternatively, if it means providing equal assistance to all income groups, or more assistance to lower income groups, coupon distribution in proportion to automobile ownership may not be appropriate. In the extreme, coupon rationing could be used explicitly to distribute income to lower income groups or to those with special needs. **(Think chaos!)**

Gasoline tax and rebate. If the rebate were allocated in the same way as coupons, the distribution of income would be similar to that under gasoline rationing. However, the distribution of money would make the implicit redistribution of income more evident to the public. Distributing coupons to vehicle owners may be perceived as fair, whereas the equivalent distribution of money may not. There is also the question of when the rebates are provided. If infrequently, there will be cash flow problems for the less well off. **(Think chaos!)**

General rebate. Uncontrolled prices would be perceived by the public as inequitable, and money would be openly transferred from consumers to oil companies. A general rebate would only partially compensate for this transfer, although the size of the rebate could be increased by adding an emergency surcharge to a windfall profits tax, which is inherent to this approach. If the rebate is distributed to all citizens, some groups will argue that others are receiving too much. Thus, in terms of the distribution of income, the fairness of this program depends on the tax rate and the structure of the rebate mechanism.

7. Practical Problems

Price and allocation controls. A major practical problem with this approach is that a comprehensive price control and allocation system would have to be imposed. Each oil company would be required to submit detailed information to federal agencies, and each refiner would have to report all oil purchases so that the government could determine the net entitlement obligations for each refiner. Oil companies would be required to maintain records of all transactions to enable government agencies to conduct audits, which would require a large federal workforce. Initiating the program would take at least several months, and even if a rationing program was developed and kept intact in standby status, changing circumstances could make any control system obsolete. **(Think chaos!)**

Gasoline queues have proved to be socially divisive in relatively small oil shortfalls, but may be less divisive in a clear national emergency. Price and allocation controls without a system of end-use allocation would result in very long lines, which would require increased security costs. Further, once in place, these controls may not be easily removed. The U.S. history of oil regulations might be a guide: The emergency controls program enacted in 1971 and 1973 did not end until 1981. In the future, as in the past, many consumers and oil companies might oppose removal of price and allocation controls after shortages ease.

Domestic price controls would allow oil-exporting nations to raise prices without the reduction in demand that would otherwise accompany such a price increase. Under price and allocation controls, the true price to the consumer is the sum of the controlled price and one of the following: The cost of waiting in line, the size of the tax or the value of the coupon. Since the real price determines consumption, oil-exporting nations could increase the price, thus reducing the length of the lines, the size of the tax, or the value of the coupon, without affecting the quantity of oil consumed in the U.S. Without controls, however, a price increase would raise the true cost of oil and result in a reduction in demand.

Coupon gasoline rationing. A major problem with this approach is that an oil price and allocation controls system would have to be imposed. Moreover, a rationing system would, in effect, create a new currency and would entail the creation and operation of a massive system parallel to the monetary system to create, disperse, transfer, and, eventually, return coupons or electronic credits to the government. On the other hand, price controls may significantly simplify the task of managing monetary and fiscal policies.

Much of the administrative cost and time required in the preparation and operation of rationing would stem from the employment and training of large numbers of government workers. Costs to the private sector would also be high. Another problem is that the information in the national motor vehicle registration file would be obsolete; the error rate could be very significant. Millions of U.S. automobiles change ownership each year, and coupons or electronic credits might be sent to previous owners of used cars while current owners receive none. **(Think chaos!)**

Gasoline tax and rebate. A major practical problem would be the imposition of a price control system and, in addition, a system of gasoline taxes and rebates would have to be legislated and implemented. Because money would be used instead of coupons or credits, many existing transfer mechanisms might be used; for example, the existing excise tax on gasoline could be raised to the desired level. Some major adjustments may have to be made to control inventory profits. However, only with considerable effort and public and private expense could existing government mechanisms such as income tax withholding, veterans' benefits, low-income energy assistance, assistance payments, or other methods be used to distribute the rebates. If rebates were distributed on the basis of motor vehicle ownership, information on motor vehicle registrations would have to be maintained. If they were distributed by adjusting income tax withholding rates, tax credits might be required for automobile owners whose rebates exceeded their tax liability. Procedures to deal with non-taxpaying or unemployed vehicle owners would be required. Because of the large income transfers, strong incentives to cheat would exist. Additional federal employees would therefore be required to monitor compliance. **(Think chaos)**

An excise tax could be set at a level that equates supply and demand; however, setting the tax at such a level is a very difficult task, and the macroeconomic consequences of mistakes could be severe. It is not clear that the government would be able to determine the correct price, for there are no precise indicators of market equilibrium, which is likely to change as shortages increase year after year. After setting the initial tax, the government would have to adjust the tax on a weekly or monthly basis as crude oil supplies and prices changed and as demand became more elastic with time. The inherent uncertainty in estimating the actual level of the tax would make accommodating fiscal and monetary policy very difficult. Further, this system would be difficult to dismantle. Rebate recipients who used less gasoline than average would not want to give up their rebates, and if history is any guide, price controls would not be easily removed.

General rebate. Very large interruptions may strain the ability of market mechanisms to function effectively. On the other hand, a major advantage of this approach is that price and allocation regulations would not have to be imposed. No new tax mechanism would be required, but new rebate mechanisms would be needed, especially to handle the enormous revenues generated by large disruptions. With considerable effort the rebates could possibly be handled as an increment to existing programs -- for example, through increased transfer payments and refundable income tax credits. However, if rebates were made strictly per capita, the rebate mechanism would be even more difficult. Assembling a master list of all citizens for the purpose of distributing rebates might be construed as an unprecedented invasion of privacy, especially since any approach would have to be brought to a state of readiness in advance of an oil emergency. If the income tax system were used, many people who do not now file income tax returns would need to do so to receive the rebate. If the oil shortage was large enough to drive prices up to the level at which the rebate would be greater than many families' withholding liability, refundable credits or a combination of withholding reductions and sales or payroll tax reductions may be required.

Procedures would have to be established to deal with hardship cases, exceptional needs of medical patients, low-income users of fuel oil, and related cases. However, the rebate program should be designed carefully to avoid measures that encourage oil use; for example, it should not reward homeowners who continue to consume home-heating oil at pre-shortage levels. All oil should be priced at its replacement value; the rebate should assist those in jeopardy of suffering and should generally restore purchasing power to the economy. The time required to implement an emergency allocation approach is also important. If, as expected, world oil prices rise rapidly at the outset of the decline in world oil production, approaches that are difficult and time-consuming to implement may be less useful; demand reductions under the general rebate approach would be immediate. **(Think chaos)**

8. Overall Evaluation

The rebate approach allows individual firms and consumers the most flexibility in adapting to the oil shortfall and provides the greatest incentive for increased domestic oil production and storage. The coupon rationing and tax-rebate approaches allow efficient allocation of gasoline, assuming the government does its job perfectly. If the burden of the oil shortfall is placed primarily on gasoline, these two approaches are less efficient in microeconomic terms than the general rebate approach. The price and allocation approach is the most inefficient in that allocations are based on historical usage or queuing, both of which impose enormous economic and social costs.

None of the approaches mitigate the supply side macroeconomic costs -- higher prices and recession -- associated with declining world oil production. The true price of oil, whether measured in terms of queues, coupons, or dollars, will be higher. Emergency allocation approaches and monetary and fiscal policies can, however, affect the demand side macroeconomic costs. The sudden, massive movement of funds into the oil market could sharply reduce output in non-oil sectors of the economy. This oil price drag is associated most dramatically with the gasoline tax and rebate and the general rebate approaches. Further, if an allocation approach can somehow exclude oil price increases from the CPI, labor agreements, contracts, and government entitlement programs that are indexed to the CPI will not escalate as rapidly as if oil price increases were included. Price controls keep oil price increases out of the CPI by requiring payment in terms of time spent waiting in gasoline lines, while rationing keeps gasoline price increases out of the CPI by creating a second currency.

While unquestionably an important criterion, equity is to a large degree a matter of perception. None of the four approaches will be perceived as fair to all groups. The general rebate approach may be perceived as the least equitable, since it allows prices to rise and enables oil companies to charge what the market will bear. Wealth will be transferred from oil consumers to oil producers, both domestic and foreign. Even if a windfall profits tax is enacted to capture most of the windfall, and even if all revenues are rebated to consumers, the public's perception will likely be one of oil companies making money at the expense of consumers. To a lesser degree, the gasoline tax and rebate approach would likely be perceived as unfair because it involves an explicit tax on consumer products. Paradoxically, the gasoline rationing approach, which is as fair as the tax and rebate approach, may be perceived as the most equitable means of allocating gasoline supplies. Even though the two approaches would probably lead to similar distributions of available gasoline supplies, the possession of a coupon confers a "right" to a gallon of gasoline in a way that currency does not. Even the coupon gasoline rationing approach will be perceived by some -- those who do not own automobiles, for example -- to be unfair. The arbitrary nature of the first-come, first-served gasoline lines resulting from this approach cannot be perceived as fair for any extended period of time.

The nature and magnitude of practical problems associated with each approach are important considerations. Three of the approaches -- price controls, rationing, and gasoline tax and rebate -- would require government to implement oil price controls. The imposition of these controls with the attendant entitlements program simultaneously with the imposition of gasoline rationing could strain some government resources. Regarding time to implement, the ability of the general rebate approach to allocate oil and oil product supplies quickly appears to give it significant advantages over other options.

The four approaches require different amounts of information on which to base decisions. The rationing approach requires projections of the volumes of gasoline available several months in advance. The gasoline tax and rebate approach requires estimates of the size of the tax necessary to equate demand with supply; such information is not now available and is not likely to be reliable even if collected. The general rebate approach requires relatively less data for the decisions required. Finally, the ease of dismantling an allocation system following a disruption must be considered. Approaches requiring any form of price controls may prove more difficult to phase out than ones that do not.

Finally, it should be abundantly clear from the foregoing that rationing is easy to say and extraordinarily difficult to implement. There is no rationing approach that stands out as simple and fair. In all cases, the required government organization to implement and manage any of the approaches will be very significant. Government bureaucracies will grow and politics will surely create special preferences. **(Think chaos!)**

C. Carpooling

Carpooling represents a significant and rapid means of reducing the consumption of gasoline and other petroleum products. The shared use of vehicles by multiple occupants applies not only to people commuting to and from work but also to a range of other people-moving activities, such as shopping, athletic events, religious events, social events, etc.

Considerable progress has been made in supporting carpooling, but it developed more from the discomfort of traffic congestion than from the desire to lower gasoline consumption. Businesses have given preferential treatment to employees who car-pool by providing subsidies, operating company van-pools, providing free parking, etc. The origin of company subsidies has sometimes been government tax treatment or programs to relieve traffic congestion. One effective government program that encouraged car-pooling is the designation of roadways as high-occupancy vehicle (HOV) lanes, reserved during certain hours for vehicles with multiple occupants.

The most wasteful practice is the single-driver trip to and from work. Consider a what-if situation. In the U.S. in 2007, light-duty transportation vehicles accounted for about 9 million barrels per day of petroleum consumption. Twenty-seven percent of the miles driven were to and from work. Therefore, around 2.4 million barrels a day were used to get to and from work. During those trips there was a weighted occupancy rate of 1.14 persons per vehicle mile. If the occupancy rate was increased 50% to 1.71 persons per vehicle mile, the total miles driven would be decreased by 50%, resulting in a decrease in petroleum consumption of up to roughly a million barrels per day of oil consumption.

While mass transportation is growing in some urban areas in the U.S., it only accounts for about five percent of trips to work. Private vehicle trips to work in a carpool account for roughly 11 percent of trips, while single-driver trips account for three-quarters of all trips. If carpooling were expanded, significant reductions in gasoline demand would result. Indirect benefits would accrue also, including a decrease in traffic congestion, a decrease in vehicle emissions, and more money in people's pockets for discretionary spending.

Significant increases in average vehicle occupancy across the spectrum of activities in the U.S. alone have been estimated to have the potential of around a million barrels per day of savings. Elsewhere in the world, significant savings are also possible, but the savings would be strongly influenced by the existing use of public transportation and variations in population densities.

In addition to individuals voluntarily changing their driving habits, organizations can stimulate carpooling by the following:

- Providing flexible work hours that allow easier work matching among employees
- Subsidizing carpool vehicles and pool drivers
- Providing preferential parking for carpools
- Providing or supporting rideshare matching services to help people find convenient carpools

Governments could:

- Offer direct financial incentives for car pools

- Provide greater tax savings for people who carpool
- Increase the availability and use of designated HOV lanes
- Create and encourage ad hoc carpool pick-up and drop-off areas
- Mandate trip reduction programs at large companies
- Provide rideshare matching programs

Government mandating of carpooling would not be an easy task, because of the following:

- Not everyone is located in places where carpooling is practical, e.g., rural areas, so a simple mandate would be inherently unfair to many.
- Employment patterns will change as oil-decline-induced recession causes shifts in business activities and commuting patterns.
- Monitoring and enforcing mandated carpooling will be both expensive and time consuming for law enforcement.

In conclusion, carpooling is a ready means for reducing oil product consumption. The high cost of fuels and fuel shortages will provide a natural impetus for people to carpool without government mandate. Whether governments decide to intercede, and to what degree, is not a simple question.

D. Telecommuting

Telecommuting is a work arrangement where employees work from home or some other work site, physically separated from the normal work place for part of the workday or part of the work week. Telecommuting is feasible where an employee's work activities can be performed remotely and is facilitated by computers, the Internet, telephones, and remote interactions via video links. By reducing or eliminating the need to travel, gasoline consumption can be reduced.

Telecommuting is a growing trend in many parts of the world, especially as a result of the Covid pandemic. When world oil production goes into decline, oil product prices will escalate and shortages will ensue, so that telecommuting will become even more attractive.

It's obvious that there are many types of work where employees cannot telecommute because they must interact directly with others or machinery. These include hospitals, emergency services, industrial production in factories, police and fire fighting, certain kinds of customer services, merchandize pickup and delivery, refuse pickup, etc. In many organizations, career advancement is dependent on working directly with other people in common locations. In these situations telecommuters could be at a significant disadvantage.

Beside gasoline savings, some of the other benefits of telecommuting include the following:

- Businesses require less office space (structures and real estate), electricity, maintenance, parking, etc.

- Employees save on transportation, food, clothing, potential daycare costs, etc. and decrease their potential for transportation-related accidents.

- Employees have more time with their families and for recreation.

- Reduced commuter traffic decreases the cost of traffic congestion and the cost of road construction and maintenance, and reduces the amount of vehicle emissions.

In the U.S., telecommuting gained substantial government recognition in 1996, when the Clean Air Act required companies with over 100 employees to encourage carpools, telecommuting, and other practices in order to reduce air emissions. The U.S. Federal Government encourages telecommuting as follows:

In 2004 Congress encouraged Federal Agencies to provide telecommuting options to eligible employees and threatened to withhold agency funding where successful programs were not put in place.

The U.S. General Services Administration had over 10 percent of its workforce several years ago telecommuting and now has a goal to reach 50 percent.

The National Science Foundation announced that approximately one-third of its employees participated in telework regularly, which offers other national security-related benefits.

Around the world significant telecommuting opportunities exist. Organizations can encourage greater participation on their own and governments working with employers can develop additional opportunities. The degree to which governments institute mandates is an open question.

One caution: The viability of telecommuting depends on maintaining the integrity of the electric power system, the phone system, and the Internet. Severe damage to any of those systems by extreme weather or terrorism would do significant economic harm, so related vulnerabilities would need to be monitored and minimized.

Appendix X. The Best That Physical Mitigation Can Provide

Note: The following is a shortened and annotated version of the 2005 study that we (Hirsch, Bezdek, and Wendling) did for the National Energy Technology Laboratory (NETL), a unit of the U.S. Department of Energy. Our report was entitled "Peaking of World Oil Production: Impacts, Mitigation and Risk Management." In the following I show the pattern that we adopted with some explanation as to how we developed our final result, which we felt to be a useful, rough picture of what might be done with physical mitigation.

A. Background

1. Introduction

At the time that our study was completed in 2005, it was considered so explosive that it was published without an official publication number, an unusual occurrence. The study developed scenarios for the most that an instantaneous worldwide mitigation crash program might accomplish. Basic assumptions in that study were that 1) No country would attempt to horde oil or key materials; 2) No oil-induced wars would occur; and 3) technology would be freely shared across borders. In the real world these are impossible, but the approach lays a starting point from which reality can be interjected.

Because the onset of the decline of world oil production ("peak oil") was very much a controversial subject, not taken seriously by governments in 2005, significant physical mitigation might not begin before peak oil becomes obvious.

In the study, it was assumed that world oil production might follow the peaked conventional oil production pattern that occurred in the U.S. Lower 48 states.

Because our 2005 rough analysis is believed to still provide useful insights, aspects are discussed here, along with rationales and results. An update of the study with the benefit of ensuing events would be useful, especially if likely real-world constraints are factored in.

2. World Oil Production Decline Rates

There are a number of clues as to the likely decline rate of world oil production. First, there are the rates that have occurred in very large oil producing regions that are well past maximum production. Second, there are rates estimated by various analysts. Third, there is the impact of exporter country policies, which could result in decline rates higher than what is physically possible (exporter withholding). Some decline rate estimates are as follows:

Actual U.S. Lower 48 States after 1970................................ 2%

Actual Europe after its plateau production...........................5-6%

Estimate from the National Petroleum Council....................4-7%

Overall range..2-7%

Human factors can and often do overshadow what nature provides. For example, faster decline rates in various countries can result from poor oil field management, inadequate investments, and poor or selfish government decision-making.

Oil field investment rates emerged as an issue in 2006, when the IEA Executive Director discussed the subject: "(The IEA) World Energy Outlook (WEO) 2006 identifies under-investment in new energy supply as a real risk." To quench the world's thirst for energy, the IEA Reference Scenario projections called for a cumulative investment in energy-supply infrastructure of over $20 trillion in real terms over 2005-2030, substantially more than was estimated to be spent as of 2006. [The situation has worsened in the subsequent decade.]

"In interviews associated with the WEO 2006 release, IEA executives further affirmed the impending difficulties: 'This energy future is not only unsustainable, it is doomed to failure,' because of, "underinvestment in basic energy infrastructure ,... In short, we are on course for an energy system that will evolve from crisis to crisis."

Two cases bracket future world oil production decline:

I. BEST CASE: A number of years of relatively flat world oil production before the onset of a decline rate of 2% per year.

II. WORST CASE: A few years of relatively flat world oil production followed by the onset of a decline rate of 4% per year.

Remember from our earlier discussion that world oil production decline rates will increase over time as more and more oil fields go into decline; growing world oil production decline rates will be unavoidable.

3. Elements of a World Oil Production Decline Analysis

A major challenge in performing a meaningful analysis of the best-case mitigation of world oil production decline is the selection of the most important variables, while making reasonable assumptions on other important factors. An exact analysis was and is impossible, because of the multitude of complexities and unknowns, but it's possible to develop rough estimates, and in the process, learn a great deal about the challenges.

Some readers may find the use of approximations and estimates unsatisfying. It can be said in response that rough estimate calculations have frequently proven robust in approximating complex situations in engineering and economics. In fact, the unknowns are such that finer-scale calculations are simply not worth the much greater effort.

4. The Delayed Wedge Approximation

In the 2005 study for DOE a "delayed wedge" method was used to approximate the scale and growth of each mitigation option, because that method captured key elements of how things work, and it simplified the analysis without loss of critical insights. Wedges were composed of two parts. The first is the startup phase that occurs before the appearance of tangible results. For example, in the case of the construction of Coal-To-Liquids (CTL) plants, a period of time is required to design, plan, and build the plants before they can begin to produce useful liquid fuels.

After the startup phase, wedges are assumed to grow linearly with time to reflect the liquid fuel savings or liquid fuels production that accumulate. The delayed wedge pattern is shown in Figure AX-1, where the horizontal axis is time and the vertical axis is market penetration, measured in millions of barrels per day of savings, as in the case of efficient vehicles, or liquid fuels production, as in the case of CTL plants. A 20-year period was used because major impacts inescapably require a long time. The further out in time that approximations go, the less accurate they'll be. While these limitations may seem troublesome to those not used to developing rough estimates, remember that the purpose was to identify major trends, which the formalism facilitates.

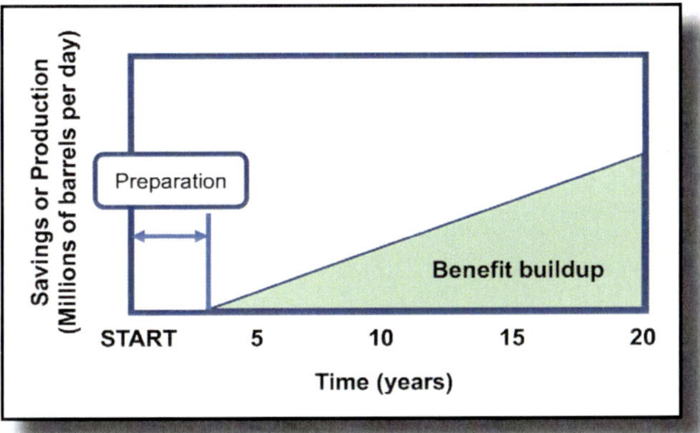

Figure AX-1. Delayed wedge approximation for various mitigation options.

The criteria used for the selection of the most viable liquid-fuel physical mitigation options were as follows:

- The option must be capable of contributions on a massive scale -- millions to tens of millions of barrels per day worldwide within a decade or two.
- The option must involve technology that is now commercial or near commercial. Technologies still in the research and development stage don't qualify until they are demonstrated on a large scale.
- The option must either save or produce liquid fuels suitable for widespread use in existing capital stock, e.g. cars, trucks, buses, airplanes, ships, industries, etc.
- Fuel production technologies must be liquid fuel efficient, which means the volumes of liquid fuels produced must be many times greater than the liquid fuels consumed in construction and operation of related facilities. Producing just a little bit more liquid fuels than are consumed in a process is a waste of time and effort.

- The option must be environmentally reasonable but not necessarily ultraclean. Every energy option has shortcomings. Compromises will be essential, because short-term human viability will be at stake.

- Feedstock resources (heavy oil, coal, natural gas, etc.) must be plentiful somewhere in the world. If not, millions to tens of millions of barrels of liquid fuels will not be possible.

Technologies that produce or save electricity in the near-term are not of interest for solving the problem of a liquid fuels shortage. However, electricity production options can be extremely valuable in their own right and will be of great value in the long term, when greater electrification of the transportation sector is developed.

5. The Options

For the purposes of rough estimates of what worldwide crash program liquid-fuel mitigation might achieve, the following were considered:

- Fuel efficient or electric transportation,
- Heavy Oil/Oil sands,
- Coal-to Liquids (CTL),
- Enhanced Oil Recovery (EOR) and
- Gas-to-Liquids (GTL)

Before discussing each in some detail, here are some options that were ignored, along with the reasoning:

Shale Oil: While the world has a huge resource of shale oil (different from oil shale) that could be processed into substitute liquid fuels, related production technologies were not ready for deployment in 2005, and there are water shortages at the most promising sites.

Biomass-to-Liquids: Ethanol from corn is currently utilized in the U.S. transportation market, because of government mandates and subsidies. Corn-ethanol is not commercially viable on its own and would disappear if it had to fairly compete with other options. Furthermore, it may not even be energy positive; the technical literature in this case is ambiguous. Biodiesel and cellulosic conversion are not yet commercially viable. Continuing research and development may change the biomass outlook, but so far, no biomass options meet the criteria.

Fuel Switching in the Electric Power Sector: As a result of the oil shocks of the 1970s, the U.S. and other nations switched most of their oil-fired electric power production to other fuels. However, the switch was not total, so opportunities for switching electric power generation away from liquid fuels certainly exist. For significant world scale impact, alternate-fueled, large electric power generation facilities would have to be retrofitted or constructed from scratch on a massive scale to provide makeup for the loss of the oil-fired electric power production. There is no way of estimating how fast such conversions might happen, because of the different economic circumstances in each country where the opportunities exist.

Nuclear Power, Wind and Photovoltaics: These technologies produce electric power, which is not a near-term substitute in transportation equipment that's built to operate on liquid fuels.

Electrified Railways: The majority of railway power in the U.S. is provided by diesel fuel, which opens up the potential for electrification. However, such an undertaking would require the construction of new electric power plants, transmission lines, and electric locomotives. Since existing diesel locomotives use electric drive, some retrofits are conceivable. However, because diesel fuel use in trains was only about 0.3 million barrels per day in the U.S., electrification of U.S trains would not have a major impact on total U.S. liquid fuel consumption. Elsewhere in the world, railway electrification is already widespread, so the overall world impact would likely be modest, which is not to deny its utility.

Building Heating: Worldwide, a significant number of homes and commercial buildings are warmed by heating oil or Liquefied Petroleum Gas (LPG). In principle these facilities could be switched to natural gas or electric heating, freeing up liquid fuels for transportation. Analysis of this path is complicated and dependent on a large number of individual and national circumstances. Keep in mind that switching on a large scale would require the construction of compensating natural gas and/or electric power source facilities and delivery infrastructure, which cannot happen quickly on the scale of world oil physical mitigation. Directionally, these conversions could be useful liquid fuels savers. However, per-facility costs are not trivial and are up to the property owner, who is often unable to afford the conversion.

Miscellaneous Activities: Many smaller scale efficiencies and liquids production options will happen. A wedge for "all other" could have been added, but it would have been strictly speculative. Furthermore, no matter what the new technologies, the magnitude of their contributions would be much less than the technologies that were considered, because of the inherently huge scale of required physical mitigation.

More Efficient Vehicles as a Mitigation Pathway

Automobiles and light trucks – Light Duty Vehicles (LDVs) -- are large consumers of gasoline worldwide. However, Hybrid Electric Vehicles (HEVs), Plug-in Hybrid Electric Vehicles (PHEVs), and Electric Vehicles (EVs) are a significant fraction of new vehicles in the U.S. and Europe.

Retooling the automobile industry to massively manufacture high mileage and electric vehicles is assumed to require three years from the decision to go ahead, because of the time needed to modify production plants and for suppliers to implement the changes needed to produce vehicle components. That estimate was in line with experience in the industry, but it's possible that the delay could be somewhat reduced. [Note that this situation is improved at the writing of this book.]

The estimate for the contributions to world fuel saving from more restrictive fuel economy standards is shown in Figure AX-2.

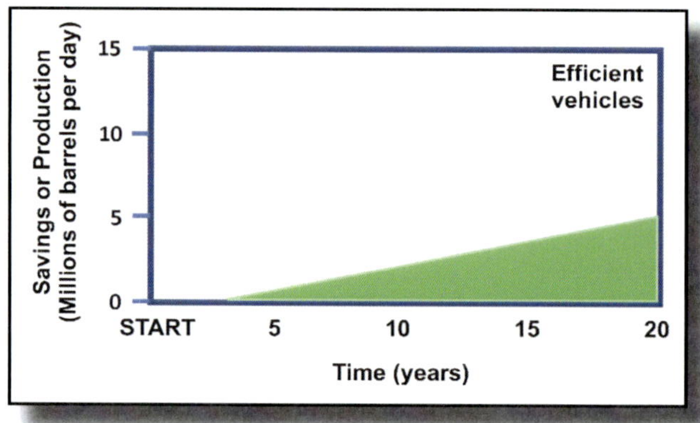

Figure AX-2. Worldwide fuel savings from much higher fuel efficient, light duty vehicles. The full wave of new vehicles is assumed to enter the market in three years after a decision to massively retool.

Heavy Oil & Oil Sands as a Mitigation Pathway

Because of their high-viscosity (goopiness), many oils do not readily flow from underground reservoirs. Such oils are called "heavy," and they often contain significant levels of impurities, which must be removed by extensive refining. Still, there are large quantities of these oils in the world, and while they're often more expensive to produce and refine than light oils, they can be economic at higher oil prices. The largest heavy oil deposits exist in Canada and Venezuela, with smaller quantities in Russia, Europe, and the U.S. Production of heavy oils has been ongoing for a very long time. While there are issues related to their environmental impacts, these issues can be managed in an oil-starved world, because oil shortages and related economic difficulties will impact people's priorities.

The assumptions for a crash program of Canadian oil sands were as follows:

- Accelerated production might begin three years after a decision to proceed with a crash program, based on the fact that the country already has significant production underway.

- A crash program expansion of Canada's previous plans assumed an acceleration by a factor of two.

- Venezuela has very large heavy oil reserves, but national chaos and mismanagement have dramatically reduced its production in recent years. Thus, any Venezuelan forecasts are problematic, because no one knows if and when the political situation will change.

The incremental crash program heavy oil / oil sands production resulting from our assumptions is shown in Figure AX-3.

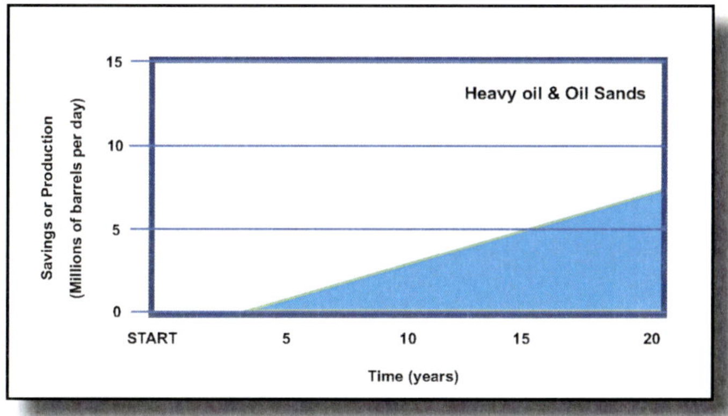

Figure AX-3. Estimated crash program of heavy oil and oil sands production from Canada and Venezuela. The estimates for Canada are aggressive, while the estimates for Venezuela assume a dramatic improvement in the country's political and investment environment.

Coal-To-Liquids as a Mitigation Pathway

Clean hydrocarbon fuels can be made from coal. The favored coal-to-liquids (CTL) process involves heating coal to very high temperatures to break the coal down into various gases. Impurities are removed, and liquid fuels and other products are created in the well-established Fisher-Tropsch (F-T) synthesis process.

A number of CTL plants were built and operated in Germany during World War II. Subsequently, the Sasol Company in South Africa built two, large-scale CTL plants under normal business conditions. A third facility -- Sasol Three -- was designed and constructed on a crash basis in response to a drop in oil imports resulting from the Iranian revolution in 1979. Sasol Three was completed in roughly three years, as essentially a duplicate of Sasol Two on the same site. We believed that the Sasol Three experience represents the very best that might be accomplished in a twenty-first century crash program to build coal liquefaction plants. The South African government made a quick decision to replicate an existing plant near its other plants without the delays associated with site selection, environmental reviews, public comment periods, etc. Experienced personnel were readily available, and there were a number of manufacturers capable of providing the required heavy process vessels, pumps, and other auxiliary equipment.

A coal liquefaction crash program assumes that coal liquefaction plants would begin operation four years after a decision to proceed. A plant size of 100,000 barrels per day of finished, refined products was assumed, and five such plants could be started each year.

The CTL crash program buildup is shown in Figure AX-4.

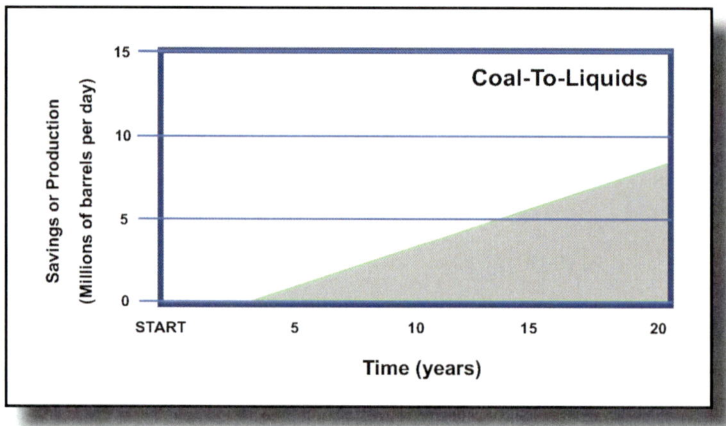

Figure AX-4. Worldwide fuel production from Coal-to-Liquids plants. CTLs begin operating after 4 years and reach a total production of 8 million barrels per day by year twenty.

Enhanced Oil Recovery as a Mitigation Pathway

Management of oil fields over their multi-decade lifetimes is influenced by an array of factors, including 1) expected future oil prices; 2) field production history, geology, and status; 3) cost and character of production-enhancing technologies; 4) the financial condition of the operator; 5) political and environmental circumstances, 6) an operator's other investment opportunities, etc.

When the decline of world oil production happens, oil prices will rise dramatically, and there is sure to be an explosion of interest in enhanced oil recovery (EOR). EOR is usually initiated after primary and secondary recovery has yielded most of what they can provide. Primary production is the process by which oil easily flows to the surface due to natural pressure in the oil reservoir. Secondary recovery involves the injection of water to force additional oil to the surface.

EOR has been in use for decades, particularly in the United States. In West Texas and eastern New Mexico EOR as of 2005 was producing about 225,000 barrels per day of incremental oil, using the injection of carbon dioxide (CO_2), which adds pressure and acts as a solvent to move remaining oil.

Thermal recovery uses heat in the form of steam to coax heavy oil out of the ground. It's used primarily in California. The Energy Department told us that about 60% of U.S. EOR is due to CO2 injection and 40% is due to thermal.

As notionally shown in Figure AX-5, CO_2 flooding can increase oil recovery by roughly 7-15 percent of the original oil in place. However, the additional oil resulting from CO2 flooding does not flow quickly; it can take a decade or more to capture most of what's possible.

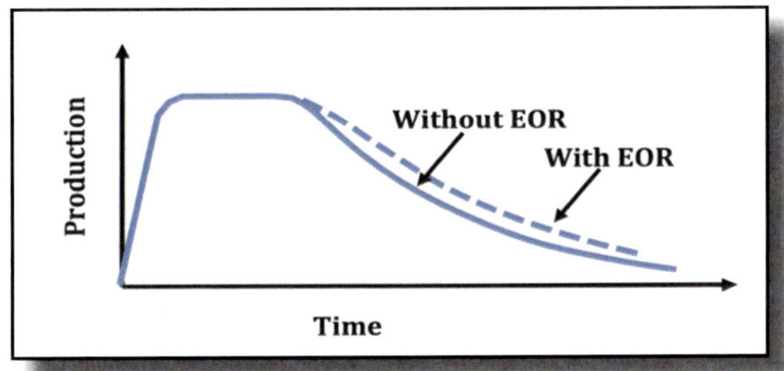

Figure AX-5. Increased production due to Enhanced Oil Recovery (EOR).

A crash EOR program will not begin to show enhanced production for 5 years after decisions to go ahead, because of the difficulties of procuring CO_2, moving it to candidate oil fields, drilling new wells, and allowing some time for increased oil flows to appear. World oil production enhancement due to such a crash effort was assumed to increase world oil production by roughly 3 percent after 10 years. This is because even under a crash program, not all oil fields are good candidates for EOR.

The estimated world EOR results estimates are shown in Figure AX-6.

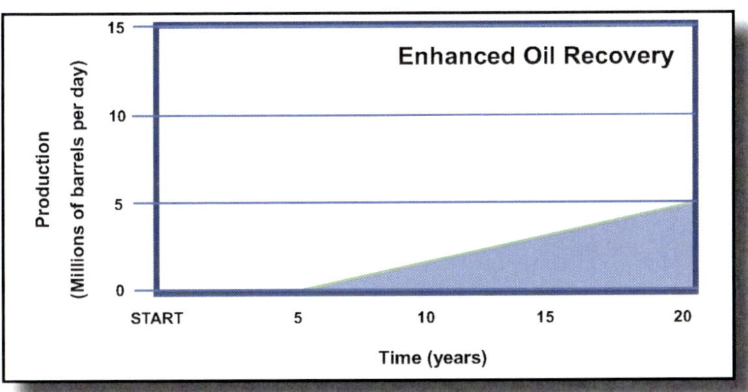

Figure AX-6. Worldwide Enhanced Oil Recovery. We assumed that this increment is achieved primarily by CO2 injection. It takes longer to begin incremental crash program EOR because of the difficulties of procuring and transporting CO2 and the drilling of needed new wells.

Gas-To-Liquids as a Mitigation Pathway

There are many large natural gas fields around the world that are isolated from gas-consuming markets. Significant quantities of this "stranded gas" are being produced, the gas frozen into a liquid, and transported to various markets in refrigerated, pressurized ships as liquefied natural gas (LNG). In a world short of liquid fuels, another possible option is to convert stranded natural gas into clean liquid fuels.

Gas-To-Liquids (GTL) processing is similar to CTL, and GTL technologies are well advanced. Developing large GTL projects involves a combination of drilling wells to produce natural gas, along with constructing large conversion facilities, which resemble very large oil refineries.

In a crash program to mitigate growing shortages of oil, GTL plants might be built in a number of countries that have large uncommitted reserves of natural gas. Once operational, GTL liquid fuels can be moved to markets around the world by conventional oil product tankers.

Worldwide GTL crash program contributions assumed a startup delay of three years before new GTL plants might come into operation. Then a crash program might yield the 1.0 MM bpd in 5 years. The resultant contributions are shown in Figure AX-7.

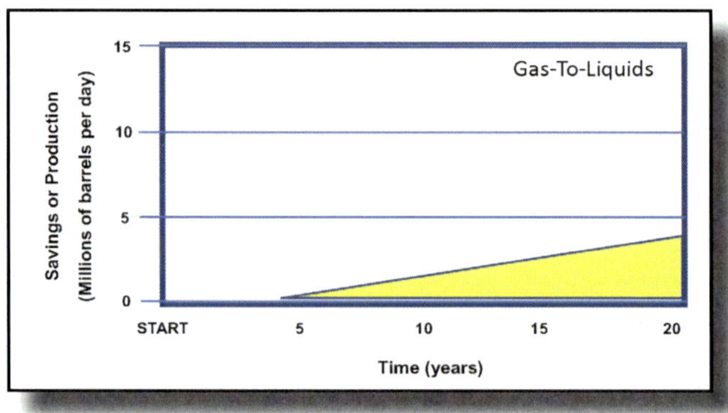

Figure AX-7. Worldwide GTL liquid fuels production as a result of a crash program.

Results of Crash Program Physical Mitigation

Review of the Purpose

The purpose of the preceding was to identify the technologies most likely to help to mitigate world oil production decline and to roughly estimate the possible contributions of each. The results provide an estimate of the best that might be accomplished, because the estimates necessarily involve making a number of judgments on what might be accomplished with each option under the best conditions. The task is much bigger than any single solution – "there is no silver bullet." To repeat, **the emphasis is on liquid fuel savings and production, because the problem is a liquids problem, not simply energy.**

The Sum Total of Crash Program Physical Mitigation

As a result of the foregoing, the five contributions can be added to generate an estimate of what an idealized crash program of physical mitigation might provide. The results are shown in Figure AX-8.

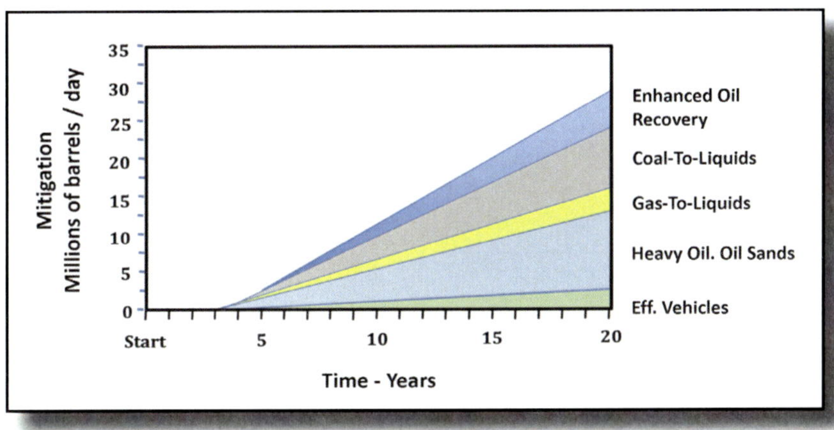

Figure AX-8. The sum total of crash program physical mitigation estimates. Contributions begin three years after a decision to go ahead and grow to roughly 30 million barrels per day after 20 years.

Crash Program Mitigation for 2% & 4% Decline Rates

Earlier, two possible world oil production decline rates of 2% and 4% were discussed. So, how might these worldwide crash programs of physical mitigation soften the oil shortages in those two cases? The results from Figure XI-9 applied to 2% and 4% decline rates are shown in Figures AX-9 and AX-10. The level of world oil production/demand has been updated to 100 million barrels per day, the early 2022 world oil consumption rate.

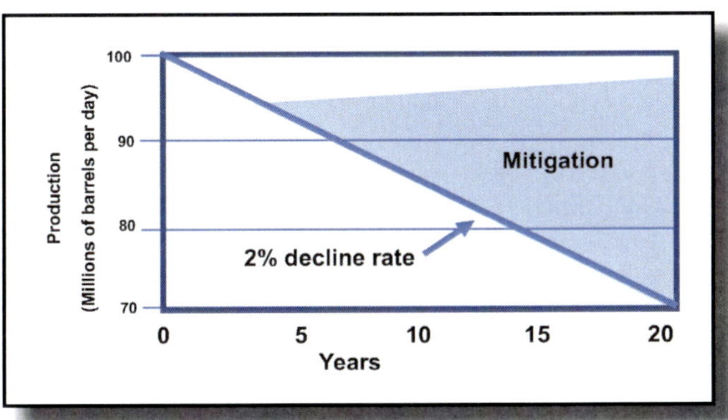

Figures AX-9. The impact of crash program mitigation on oil supply when world oil production declines at 2% per year.

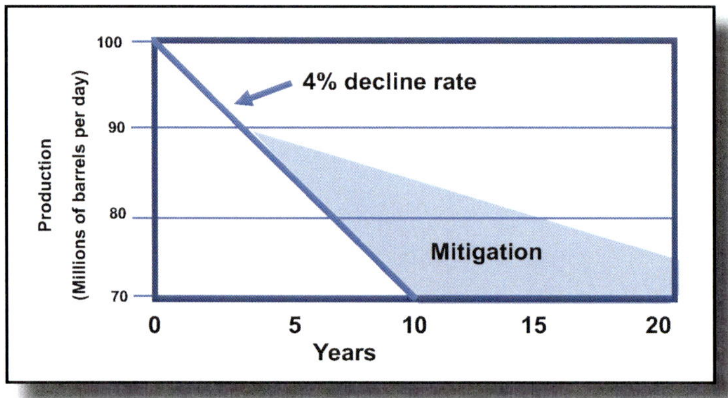

Figures AX-10. The impact of crash program mitigation on oil supply when word oil production declines at 4% per year.

Reminder: In both cases, an idealized, monumental, worldwide crash program of mitigation was modeled, something never before attempted.

In the hypothetical 2% decline rate case, the resulting world oil shortages would be extremely damaging. In the 4% case, economic damage would be very severe.

What Might a Realistic Worldwide Crash Program Look like?

The model's crash program was postulated to start instantaneously, meaning that business-as-usual would be instantly followed by full-blown implementation. **This assumption is useful for understanding the best that's possible, but in the real world, such huge changes cannot happen that quickly for a number of reasons:**

Governments will have to decide to embark on similar crash programs.

Governments will have to facilitate the rapid permitting of facilities, eliminate legal challenges to moving ahead quickly, provide incentives for industry to act aggressively, and provide financial protection for industries that undertake large mitigation risks. It is virtually certain that governments will not move quickly on such matters, because of the need to change priorities and to understand the practical options.

There is such strong sentiment among environmentalists and people concerned about climate change that there will be huge pushback until it's obvious that the decline of world oil production is causing severe hardship.

Industry will not have the capacity, infrastructure, and trained personnel to dramatically expand into a full-blown crash program overnight. Where industrial capacity does exist, it's often located in other countries, where local priorities would almost certainly restrict providing services or hardware to other countries.

The efforts envisioned will be very expensive, and the large financing requirements are unlikely to be available quickly.

Not all countries with the resources to undertake meaningful physical mitigation may do so quickly.

Any effort by governments to pick winners and losers in advance would almost certainly be disastrous, based on recent history.

Finally, there is a tendency of governments to want to exert control over big programs. If excessive, government control could severely damage what only private industry can do effectively.

Thus, a realistic implementation pathway is certain to be slower than the model postulated, more like what is shown in Figure AX-11. The result is that oil shortages will almost certainly be larger than in our idealized model.

Why not assume a more likely, slower startup crash program pathway? The answer is that the model sought the best possible response, which would be an overnight scale-up. In addition, an analysis of more likely startup pathways is an incredibly complicated job involving a number of subjective assumptions. Figure AX-11 show a notionally more likely startup, compared to our idealized step change startup.

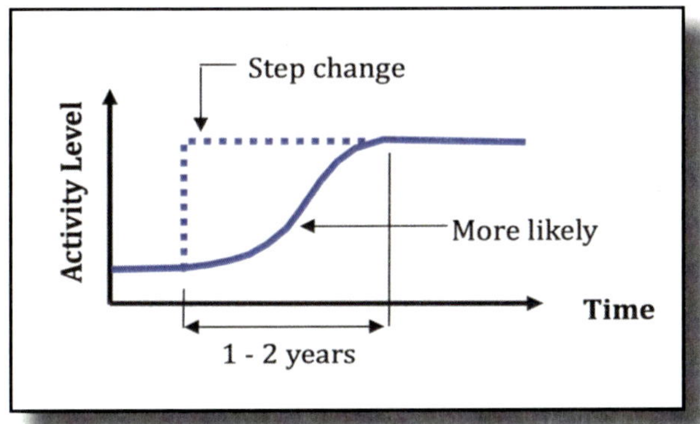

Figure AX-11. Comparison of an idealized step change pathway with a more likely pathway to change.

From these considerations, it’s clear that the mitigation results shown in Figures AX-9 and AX-10 are certain to be slowed, resulting in even more serious oil shortages.

Remember, the purpose of this exercise was to generate a general picture of what the post peak oil mitigation world might look like under two possible world oil production decline rates.

Conclusions

The conclusion is that the world is facing significant oil shortages after the onset of the decline in world oil production. **If the decline rate is a relatively modest 2%, the shortages will still be catastrophic but less onerous than if it is 4%.**

As noted, there's no question that an army of analysts could do a more detailed study. Such an exercise might be justified if the federal government was doing it a decade or more before the advent of peak oil in order to mobilize resources in advance of the event. However, since federal governments are not noted for detailed long-range planning, it's hard to envision that happening. I wish it was otherwise.

Appendix XI. What Forecasters Have Forecast

A. The Periods Covered

The following overview of what forecasters have said about future oil production is broken into two groupings. First is a sample of what people said in the period prior to 2010, a period when the subject of peak oil was hotly debated. The concern back then was the geologically-based peak – when the world oil maximum would be reached because the earth's endowment of oil would no longer support human needs.

The second period is from 2018 to the writing of this book in 2022, when concerns focused on a world underinvestment-based peak. The following is a useful sampling of opinions, but it's by no means complete. Apologies to any expert that I might not have properly included. Bold is used to highlight people and organizations quoted.

B. Prior to 2010

Rex Tillerson, Chairman and CEO of ExxonMobil in March 2009 "celebrated the earth's continued abundance of oil, noting that humans have consumed barely a third of the planet's available petroleum reserves."

Michael C. Daly, then Group Vice President, Exploration & LTR, BP Corp., in a 2007 speech said, "I believe, from what I know today, that **peak oil supply is still a long way off.**"

David Eyton, Head of Research and Technology at BP, said in Delhi in January 2009: "**we estimate that the world has already demonstrated the commercial viability of around 40 more years of conventional oil resources, 60 years of gas and 130 of coal at current consumption rates.**"

Cambridge Energy Research Associates (CERA), long optimistic, opined that "**under expected conditions, world oil supply will continue to expand until roughly 2030, after which there will be a long plateau in production.**"

Robert W. Esser, a geologist and CERA's senior consultant/director of global oil and gas resources, in September 2007 stated, "**Peak Oil theory is garbage as far as we're concerned.**"

Michael C. Lynch. Then President and Director of Global Petroleum Service, Strategic Energy & Economic Consulting, Inc., and one of the better-known resource optimists. He has a long history of challenging "peak oil". For instance: "**The prospect of an oil production peak at 100 mb/d, …, appears unlikely in my opinion, …oil production will probably pass 100 mb/d within 12-15 years.**"

The National Petroleum Council in 2007, "The capacity of the oil resource base to sustain growing production rates is uncertain. Several outlooks indicate that **increasing oil production may become a significant challenge as early as 2015.**"

The U.S. General Accountability Office did an oil supply in 2007 and stated "Most studies estimate that (world) oil production will peak sometime between now and 2040... **an imminent peak and sharp decline in oil production could cause a worldwide recession.**"

The German Energy Watch Group in its 2008 report, entitled "Crude Oil – The Supply Outlook," the group concluded: "**Peak oil is now**."

The Global Witness Foundation in a 2009 study: "This report has sought to demonstrate that the **four underlying fundamentals of oil field depletion, declining discovery rates, insufficient new projects, and increasing demand, though obvious for a long time have not been acknowledged, or acted upon…**"

The UK Industry Taskforce on Peak Oil and Energy Security in a 2008 report: "**The next five years will see us face another crunch - the oil crunch.**"

The UK Energy Research Centre observed in 2009: "A global peak is inevitable. The timing is uncertain, but the window is rapidly narrowing…**There is a significant risk of a peak before 2020**."

The U.S. Department of Defense Joint Forces Command in 2008 noted: "**By 2012, surplus oil production capacity could entirely disappear …**"

The International Energy Agency (IEA) "... warns that the credit crisis and project cancellations will lead to **no spare crude oil capacity by 2013**."

Goldman Sachs opined: ""Our analysis points towards **100% OPEC capacity utilization by 2011/12, leading to the need for demand rationing pricing.**"

Matt Simmons was the Chairman of Simmons & Co. International, and author of the book "Twilight in the Desert." He sounded **the "peak oil" alarm for a number of years**.

T. Boone Pickens, oilman and businessman developed "The Pickens Plan" to solve a number of the U.S. energy problems. In November, 2009, Pickens stated that "**the world has 'maxed out' at 85 million barrels a day of production**"

Jim Mulva, Chairman of ConocoPhillips, said he expected "it will be hard for crude supply to meet demand in the years ahead, with **output possibly peaking below 100 million barrels per day."**

David O'Reilly, Chevron Corp. Chairman and Chief Executive, warned of **a potential oil supply shortfall midway through the next decade** that could potentially trigger a substantial increase in prices.

John Hess, Chairman and CEO, Hess Corp., in, 2009: "we will have a **devastating oil crisis in the next 5-to-10 years."**

Charles Maxwell of Weeden & Co. is often called the dean of oil analysts. He recently wrote that "peak oil is **now projected for approximately 2015**."

Christophe de Margerie, CEO of the oil company Total, said that, "If we don't move [now] there will be a problem. **In two or three years it will be too late**."

Tom Petrie of Bank of America Merrill Lynch stated that **"we are at peak oil."**

Fatah Birol, Chief Economist at the **International Energy Agency (IEA)**: "IEA recently found that **most of the biggest fields have already peaked** and that the rate of decline in oil production is now running at nearly twice the pace calculated just two years ago."

Ray Leonard is CEO of the oil and gas company **Hyperdynamics Corp**.: **"Peak oil is likely within the next two-to-three years**, assuming that demand growth starts to recover."

Jose Gabrielli, the Chief Executive Officer of Petrobras, the Brazilian oil company, December 2009 stated that **"the world needs oil volumes the equivalent of one Saudi Arabia every two years."**

C. The Period 2018 -2022

Rystad Energy, a highly-respected consulting company, said **"discovered resources at all-time low for 2017."** Jan. 8, 2018.

Alan Brooks, Editor of PPHB, MUSINGS FROM THE OIL PATCH, January 9, 2018: As one money manager put it last fall: **"If Saudi Arabia is selling its oil company, what does that say about their view of the long-term future for the oil business?"**

The International Energy Agency (IEA) in the Oil and Gas Journal, March 5, 2018: "Oil production growth from US, Brazil, Canada and Norway can keep the world well supplied, more than meeting global oil demand growth through 2020, … but **more investment will be needed to boost output after that.**"

Rystad Energy in oilprice.com, March 5, 2018, "… liquid resources from mature assets grew 151 billion barrels over the last four years, which is **almost 17 percent more than the amount produced in those years.**"

Dr. Charles Hall in Peak Oil News & Message Boards. March 5, 2018: "…as a global society, we are **hurtling towards an energy crisis** that will forcefully (and likely painfully) downshift our standard of living within the lifetime of the current generation."

5 Key Takeaways From **CERAWeek. Nick Cunningham. OilPrice. com. Mar 11, 2018, "…** the real fear at this year's conference was **the possibility of a shortage of supply in the 2020s**…"

Roger Blanchard. The World Oil Supply is Infinite. Resilience.org. March 12, 2018: "In recent years, the rate of global oil discovery has been running less than 1/5th the rate of global oil consumption. It appears that **the 2017 discovery rate will end up around 1/10th of the consumption rate."**

Goldman (Saks): Oil Demand Will Continue To Soar. OilPrice.com. April 24, 2018. **Oil demand is growing, and production is falling.**

Russia's Peak Oil Production Could Be Just Three Years Away. OilPrice.com. Sep 20, 2018. Tsvetana Paraskova. **Russian Energy Minister Alexander Novak. "Russia's oil production could peak as early as in 2021** due to high taxes and costs, provided there are no benefits for exploration or tax incentives introduced**."**

The Inevitable Oil Supply Crunch. OilPrice. com. Nick Cunningham - September 27, 2018, "The warning signs are there – the industry isn't finding enough oil." That's the start of a new report from **Wood Mackenzie**, which concludes that **"a supply gap could emerge in the mid-2020s** as demand rises at a time when too few new sources of supply are coming online."

Saudi Arabia Calls The End Of Russia's Oil Prowess. OilPrice.com. Oct. 16, 2018, "Saudi Arabia has not only called the end of Russia's prominence as a global oil behemoth, but anticipates that **Russia's oil exports will have declined heavily if not disappeared within the next 19 years, Mohammed bin Salman** said in a recent interview with **Bloomberg."**

IHS Markit: Conventional oil, gas discoveries at 70-year low**.** OGJ. Oct. 2, 2019. "**Conventional oil and gas discoveries during the past 3 years are at the lowest levels in 7 decade**s and a substantial rebound is not expected, said IHS Markit in a recent report."

The State of Exploration 2019. **westwoodenergy.com. May 7, 2019. "The search for new frontier plays remains a challenge** and **the commercial success rate for frontier exploration over the last five years has been just 6%.** The continuing large gap between technical and commercial success rates in very high-risk wells points to a persistent over estimation of volumes pre-drill, with **only 15% of discoveries over the last five years in the very high-risk category being deemed commercial."**

The End of the Oil Giants, And What It Means. **SRSocco** Report. May 15, 2019 "Recently, Saudi Aramco, the world's largest oil exporter, has acknowledged that **Ghawar, the world's largest oil field, is in decline**. The news went mostly unnoticed except in the specialised media."

Saudi Aramco reveals sharp output drop at its major fields. World Oil. Javier Blas. April 2, 2019. "When Saudi Aramco on Monday published its first ever profit figures since its nationalization nearly 40 years ago, it revealed that **Ghawar is able to pump a maximum of 3.8 MMbpd—well below the more than 5 MMbpd that had become conventional wisdom in the market."**

Industry Set for Worst Discovery Toll Since 1946. Rigzone Staff by Andreas Exarheas. December 21, 2021. " **The oil and gas industry is on course for its worst discovery toll since 1946,** according to new **Rystad Energy** analysis."

Canadian oil output set to peak in 2032 - energy regulator. Financial Post. Nia Williams December 9, 2021. "Oil output from Canada, the world's fourth-largest producer, is **set to climb over the next decade to peak at 5.8 million barrels per day (bpd) in 2032**, the **Canada Energy Regulator** (CER) said."

Jefferies: Fully Reopened World Could See $150 Oil. Oilprice.com. Tsvetana Paraskova. Dec 01, 2021. "In a really fully reopened world, the oil price could go to a $150 dollars because the supply constraints are dramatic."

The oil system is collapsing so it may derail renewables warn **French Government Scientists.** Byline Times. Nafeez Ahmed. 20 October 2021. "A team of French government energy scientists are warning that … **in just 13 years, global oil production could enter into a terminal and exponential decline.**"

Is The World Sleepwalking Into An Oil Supply Crunch? Forbes. **Wood Mackenzie.** July 15, 2021. "Under-investment in oil supply will lead to a tight oil market later this decade. It's a narrative that's gained increasing traction as capital expenditure on upstream oil and gas has shrunk. **The supply is there to fill the gap to 2030**."

It's Too Late To Avoid A Major Oil Supply Crisis. Oilprice.com. David Messler - June 17, 2021. "Planned investment in oil supply globally falls about $600 billion short of what will be needed to meet projected demand by 2030, according to **JPMorgan Chase & Co. Pressure to deliver cash to shareholders, partly driven by worries about the long-run outlook for oil demand, has limited the industry's ability to plow money into new projects."**

Oil Major Total Sees 10 Million Bpd Supply Gap In 2025. Oilprice.com. Feb 10, 2021, **Tsvetana Paraskova. " France's supermajor Total** is warning that **the world could find itself with a shortfall of supply of 10 million barrels per day (bpd) between now and 2025,** due to continued underinvestment in the industry, the OPEC+ pact, and cracks in the U.S. shale business model."

2021 global oil and gas discoveries projected to sink to lowest level in 75 years. Rystad Energy. rystadenergy.com. December 20, 2021. "**Global oil and gas discoveries in 2021 are on track to hit their lowest full-year level in 75 years.**"

Can The World Avoid A Global Oil Supply Crunch? **Oilprice**.com. January 12, 2022. **"OPEC's spare capacity is dwindling, new discoveries are at historic lows, and banks are growing increasingly reluctant to engage with the oil and gas industry because of the rise of ESG investing."**

Analysis: When it comes to oil, the global economy is still hooked. Reuters. March 25, 2022. The world may be less dependent on oil now than it was during the energy shocks of the 1970s, **but the Ukraine conflict is stark evidence of a stubborn craving that can still disrupt economies, confound policymakers and spark political strife.**

Appendix XII. Electric Power Options

A. Introduction

Thankfully, the U.S. electric power generation system is in good shape. In 2021 EIA tells is that fuels for electric were roughly as follows:

Natural gas..................... 38%

Coal...............................22%

Nuclear...........................19%

Renewables.....................20%

Of the renewables, roughly 9% was wind; 6% was hydroelectricity (waterfalls); 3% solar cells, and the rest was geothermal and biomass (burning wood, etc.).

B. Wind and Solar Cells

Both wind and solar cells produce intermittent electric power; their performance depending on the vagaries of nature. Wind does not always blow at a constant speed, which means that wind-based electric power can vary dramatically hour-to-hour, day-to-day, month-to- month, etc. You've felt wind variations in your everyday experiences.

5 MW wind turbines in the North Sea off the coast of Belgium.

Solar cells work at their maximum around noon on clear days. When there are wispy clouds, output can drop by maybe 10%. On days with heavy clouds, output can drop 70-80%, and at night solar cell output drops to zero (no light in = no electric power out).

I've been entranced with solar cells since I was 12 years old. That's when I was given a light meter, which had a solar cell as its means for measuring light conditions to set my camera aperture. In those days light meters were external from cameras, unlike today when they're typically built in. What a truly marvelous invention! Light enters a solar cell, providing an electric current proportional to the amount of light coming in. That current registers on a meter, calibrated to show light intensity.

When I was in the federal government, I had responsibility for federal solar research among other programs. I was given a solar cell mounted in a 1" x 1" wood block, with a meter showing light intensity as a function of solar radiation level (watts per square meter). Over the decades I periodically took that little meter outside and measured solar radiation under different cloud conditions and can personally verify the levels of solar radiation that I quoted.

Solar array mounted on a rooftop

So, what are the drawbacks to wind and solar cells with their variable electric power outputs? As consumers, we expect our electric power to be available when we want it; the lights should go on when we flip a switch; the television should turn on when we hit the remote, etc. **We insist on power-on-demand. Wind and solar cells cannot provide it, so other electric power sources have to be ready to make up for the nature-limiting shortcomings of those technologies.**

The stand-alone costs of wind and solar are NOT the price of those technologies alone, because they don't provide power-on-demand. Their legitimate cost is that for the units themselves and, often, gas-turbine generators that must be paired to provide the electric service we demand. Why gas turbines? Because they can be quickly revved up on request. That reality is rarely stated, because it means that **wind and solar are not as attractive as proponents contend.**

You might ask about energy storage to level the power out of wind and solar. In most cases that must be batteries. The problem is that batteries are hellishly expensive for full-scale electric power applications. Maybe someday.

C. Nuclear Power

Nuclear power is a potentially attractive technology, when it's properly deployed and managed. However, there were instances when nuclear power accidents frightened and killed people.

The 1200 MWe Leibstadt Nuclear Power Plant in Switzerland. The reactor is housed in that globe-like building. The huge structure emitting steam is the cooling tower.

What comes to mind? There's the Chernobyl disaster of 1986. In that case the Russians built a type of nuclear reactor that was inherently unstable (there are many different types of nuclear reactors) and built them without a robust enclosure to contain any radioactivity that might leak or an explosion in an extreme situation. What happened at Chernobyl? Operators did some not-carefully-thought-out experiments and the reactor melted down, causing fires, and negatively impacting adjacent reactors. It took a long time to put out the fires and isolate the mess but not before radioactivity spread over a huge area of Ukraine and Europe. People close-in died and others were disfigured. The Russians brought this on themselves with their unstable technology, and all of nuclear power got a bad name, even though it was an irresponsible action on the part of the Russians.

Another nuclear reactor failure was the Fukushima nuclear disaster of 2011 in Japan. In that case, reactors were constructed close to the ocean to provide easy access to cooling water. Sadly, the developers didn't consider the possibility of a tsunami striking the reactor site. One did, and reactor systems failed catastrophically. One person was killed and maybe 40 were injured. Many safety features worked, but a few didn't. The reactors should never have been located where they were.

Lastly, there's the Three Mile Island (TMI) nuclear reactors near Middletown, Pennsylvania, that had an accident in 1979. The cause was a combination of a failure of the cooling water system and the inability of plant operators to understand the reactor's condition during a malfunction. This event might be considered a "successful failure." What? TMI was built according to strict U.S. Nuclear Regulatory Commission standards, and among other things, had a large, robust containment vessel that enclosed the reactor and various support equipment. The result was that the accident was almost completely contained with a very small amount of radioactivity leaking outside the system.

The good news is that a great deal of learning has gone into subsequent nuclear power plant design and construction. The bad news is that nuclear regulation may have gone too far in that it's so super careful that it's made the cost of new plants very expensive. As a result, only two new nuclear plants have come into operation recently, and only two additional reactors are under construction in the U.S. at the time this book was written.

Did you know that the French are very heavily dependent on nuclear power? France derives over 70% of its electricity from a host of nuclear reactors. Early on, the French made a decision to standardize their nuclear plants, unlike the U.S., where utilities wanted customized reactors, making regulation more difficult. France had one significant reactor accident over the decades with no one hurt. Besides some small events, their nuclear system has been remarkably accident-free.

In recent years, there's been a great deal of interest in so-called Small Modular Reactors (SMRs). These systems are much smaller than their earlier brethren to the point that they could be manufactured in factories and shipped to their intended locations, saving on the cost of building large systems on a power plant site. SMRs are still in the development stage.

Let's not forget nuclear waste, which is very dangerous stuff. The good news is that the volumes of U.S. nuclear waste are relatively small and can be buried in geologically stable places. The bad news is that decades ago a site in barren country located well outside of Las Vegas was chosen for U.S. nuclear waste burial, but it ran into intense resistance and was abandoned. The French chose to recycle its nuclear waste, reusing unused fuels and reducing waste volumes dramatically. What's left is sealed and will be buried in a designated site in the future.

Could nuclear reactors be used to produce liquid fuels? There've been some concepts kicked around, but none reached the serious stage.

D. Fusion Power

Fusion energy research has been ongoing in the federal government since the 1950s. In my opinion it's by far the most screwed up research program that I've ever witnessed. At a budget of over $650 million dollars per year, it's tragically off course from its goal of creating a practical new source of electric power. The program has never had an independent external oversight by objective technologists; all its reviews have been by people in or associated with the fusion program: "Foxes watching the hen house!" Reviews are not easy because the technology is very complicated. Nevertheless, all that needs to be considered is the complexity and cost of the particular concept that the federal fusion program is focusing on.

Who am I to criticize? I did fusion research myself and rose to lead the federal program.

When I took over in 1972 with a new team of talented personnel, we cleared out losing projects and focused on the tokamak concept for fusion plasma confinement. To explain:

Fusion is the fundamental energy source in the universe. Our very hot sun is a fusion reactor held together by gravity.

Fusion operates using plasma, a gas that's so hot that all atoms of gas are broken apart into the fundamental constituents of ions and electrons.

Tokamak is a "magnetic bottle" composed of magnetic field coils arranged in a hollow, donut-shape, where hot fusion plasma is held inside, away from material walls, so plasma can be heated to hundred-million-degree temperatures.

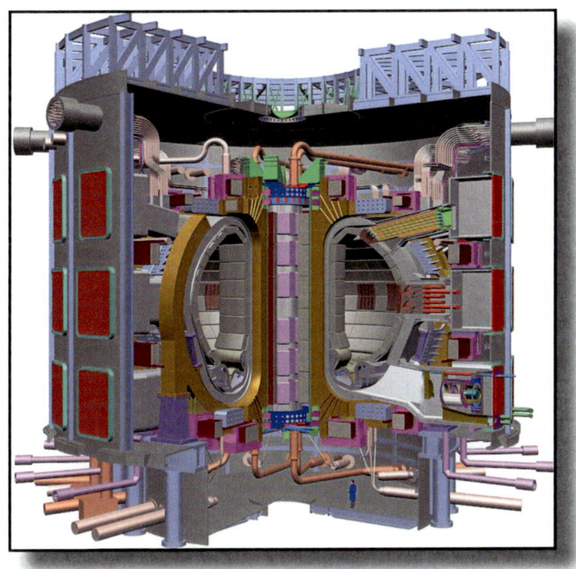

A cutaway pictorial of the internationally-funded ITER tokamak fusion reactor under construction in France. That little vertical line near the bottom on the right is a human. ITER is not being built to generate electricity. If energy conversion equipment was attached, ITER would use all of it 500 megawatts of thermal energy to just run itself.

Sound complicated? It is. The science of hot plasmas in magnetic fields turned out to be extraordinarily more difficult than experts thought 70 years ago when the program started.

There are a number of magnetic "bottles" that might hold super-hot fusion plasmas. Such systems are needed to keep the plasma away from metal walls, which would extinguish the plasma and damage the walls in the process.

When we selected the tokamak concept for emphasis back in 1973, it was the most attractive concept for making a fusion power reactor of all that were then under study. Bigger and more expensive tokamak experiments were built by us and successor government program managers.

What had appeared potentially attractive in the 1970s evolved into a system that is hugely expensive and extremely complicated. And there are still questions if it will even work properly. Rather than periodically asking engineers to evaluate what they were working on, physicists were and are happy doing good physics research; good physics is measured on a physics scale, not a practical engineering scale. The mantra of the competent engineer is to build things of practical value. The physicists took over the program after we left and stayed with the tokamak concept, even though it became an engineering nightmare - grotesque is not too strong a description.

Why didn't senior management in the Department of Energy step in and demand that the fusion program be objectively evaluated and, as I believe, revamped? Because the physics was and is so very complicated, and the physicists seemed so intent on continuing with tokamaks. The physicists have managed to keep commercially-oriented engineers away from evaluating what they're doing, even though the program is supposed to be developing a practical fusion power reactor.

Things that are wrong with the very expensive federal fusion research program are as follows

As larger and larger tokamak experiments were built, they became an engineering nightmare of complexity.

A really huge tokamak experiment was planned and made an international project. ITER is now under construction in France. It was initially supposed to cost roughly $5 billion. At that cost, the tokamak might have led to a marginally cost-effective commercial electric power source. **But the cost of ITER swelled to more than $20 billion, and it's still climbing, way too expensive to be commercially attractive.**

There are a number of possible fusion fuel cycles but the one the fusion people selected yields copious quantities of neutrons that will make reactor walls and structure "swiss cheese" weak in a matter of maybe two years. In addition, the neutrons will make the core of a tokamak power reactor highly radioactive. Read **huge volumes of radioactive waste**.

The fuels for a commercial tokamak reactor, which were thought to be widely available, are no longer so, resulting in a concept with inadequate fuel worldwide.

These things evolved slowly over time. The physicists were happy doing physics research, while the cascading engineering problems were "swept under the carpet."

Why do I go on like this about federal fusion research? Because I continue to have hopes for attractive, commercial fusion electric power, and the engineer in me clearly concludes that **tokamak research has evolved into a gigantic loser**. The citizen in me hates to see tax money wasted. The manager in me feels some responsibility for having selected tokamaks for emphasis some 40+ years ago. Hopefully, the fusion sham will be exposed soon and better directions for federal fusion research established.

Appendix XIII. The Russia-Ukraine War: A Dry, Dry Run-on Peak Oil?

A. Introduction

At this writing, the Russia-Ukraine war is weeks old. During that period, we've seen some relatively dramatic changes in a variety of commodity prices, including oil. It's far too soon to draw long-term conclusions from these events, however, **it's not too early to make some observations related to the subjects of this book.**

Worth noting: The Russian-Ukraine war was a shock to many and maybe that kind of shock is a little like what can be expected when it becomes known that world oil production has or will reach a maximum. Another point is that inflation is running rampant, and people are hurting financially. They're also looking to place blame.

Following are some notes and quotes from articles I considered meaningful, either because of who was quoted or because of their credibility. The references for the quotes are brief but should be sufficient for the reader who wants more.

B. Comments on Various Utterings

A number of commentators and politicians are "whistling in the dark" on oil pricing. They don't seem to understand the nature of the oil business, and **they seem paralyzed as far as what to do or what to recommend to bring oil prices down.**

I've heard some from both groups say that "Energy independence is the Holy Grail." People uttering such things don't understand energy, because **the current issue is oil, not "energy" for the U.S.**

Some have said "Open the spigots." This implies that the speakers think that oil companies are holding back or that producing more oil can be done quickly, both of which are wrong.

It's been said that building the Keystone pipelines would dramatically change U.S. oil prices. That is wrong. The pipeline would allow additional Canadian oil to flow to the U.S., but it wouldn't make a noticeable difference in world oil supplies. A significant amount of Canadian oil already flows to the U.S. via tank car trains and existing pipelines. Also, oil prices are established on a world-market basis, not on a U.S.-only basis.

C. Quotations From the Media

"The world may be facing one of the largest energy shocks ever," Goldman Sachs says. "The uncertainty on how this conflict and oil shortages will be resolved is unprecedented," CNN Business. March 09, 2022.

“Oil producers [are] in a ‘dire situation’ and unable to ramp up output”...U.S producers were largely expecting to keep oil production flat this year, and in the face of surging crude prices, output can’t just be ramped up,” said Oxy [Petroleum] CEO Vicki Hollub. The oil-rich Permian Basin faces significant challenges in boosting output, according to Hollub. But it’s the only shale basin in the U.S. that can increase production, she said. Among the challenges are the age of the wells, labor shortages and securing raw materials. CNBC. March 8, 2022.

“Venezuela’s cash-strapped government is eager to increase production [but] rusted and frequently damaged oil pipelines carry a fraction of the crude they once could. Rigs that once dotted the landscape in the country’s oil-rich sectors have gone quiet. Looters break into oil installations to steal everything from pumps to compressors to fencing and metal to be sold as scrap.” WSJ. March 8, 2022.

“Ukraine War Hits Farmers as Russia Cuts Fertilizer Supplies, Hurting Brazil. **Brazil is searching for new fertilizer suppliers as the war in Ukraine threatens to cut off shipments to one of the world’s breadbaskets,** with potential ripple effects on already high global food inflation. The Latin American country is the largest producer of coffee, soybeans and sugar, and the most dependent of the world’s agricultural superpowers on imported fertilizer. Brazil imports some 85% of its fertilizers and about a fifth of those imports come from Russia. The Russian trade ministry has called for a broad suspension of fertilizer exports, state news agency TASS reported Friday.” WSJ. March 5, 2022.

“Will the Russian Invasion Accelerate Peak Oil? Soon the globe could have more clean energy—or higher-priced oil. **The proliferation of climate pledges in recent years has come against the backdrop of soaring fossil fuel consumption.** The International Energy Agency estimates oil demand will surpass pre-pandemic levels this year. E&E News. March 4, 2022.

D. Climate Change Changes

“China sees biggest growth in energy and coal use since 2011. China, the world’s biggest coal burner and greenhouse gas emitter, used 5.24 billion tonnes of standard coal equivalent of energy last year, up 5.2% from 2020, the [Chinese] National Bureau of Statistics (NBS) said.” Reuters. Feb. 28, 2022.

“At global energy conference, leaders acknowledge fossil fuels not going away soon. Even Energy Secretary Jennifer Granholm said the western world is on a “war footing,” **and energy executives need to start “producing more right now.” The big story at CERA this week is that basically people recognize that carbon is still a major factor in the global energy market,” Carlyle Group Co-Chairman David Rubenstein told FOX Business. “And while there’s been more discussion on renewables over the last year than carbon, in truth, carbon energy is still here, and we still are very dependent on it. And it’s not gonna change any time soon.”** Fox Business. March 13, 2022.

“As Gas Prices Soar, Buyers Find Fewer Fuel-Efficient Car Options. **With SUV and truck sales booming in recent years, many auto makers scaled back on smaller options;**

electric-vehicles remain limited. Americans looking to offset surging gasoline prices with a more fuel-thrifty vehicle aren't likely to find much on the car lot today. The sharp rise in fuel costs, driven in large part by the war in Ukraine and related disruptions, is delivering another shock to the car business and triggering renewed focus on fuel economy after a multiyear boom in sport-utility vehicle and pickup truck sales. It also comes as new- and used-car inventory on dealership lots is at historic lows, leaving buyers with slim pickings for those looking to make a switch, according to dealers, executives and analysts." WSJ. March 13, 2022.

"**Geoengineer the Planet? More Scientists Now Say It Must Be an Option.** Human intervention with the climate system has long been viewed as an ill-advised and risky step to slow global warming. But with carbon emissions soaring, initiatives to study and develop geoengineering technologies are gaining traction as a potential last resort. Some fear that suggesting an easy fix for global warming will encourage delay in emissions reductions." YaleEnvironment360. May 29, 2019

"Rich Countries Must End Oil and Gas Production by 2034 for a Fair 1.5°C. **International Institute for Sustainable Development.March 22, 2022.**

E. World Oil Prices and Production

The average oil price in 2019 was $64.3 per barrel. **The U.S. crude oil spiked to 13-year high of $130 per barrel overnight**, then gave up most of that gain. CNBC. March 6, 2022.

"John Hess, CEO of oil and gas producer Hess Corp., said last week's announced oil release, part of **international disbursement of 60 million barrels, was too modest to affect the market.** The Biden administration release amounts to about 1.5 days of U.S. consumption." Politico. March 7, 2020

"High Oil Prices Aren't Enough To Tempt Shale Producers. America's shale industry is looking to ramp up production, but it is facing two major hurdles that could curb its trajectory. **Supply chain issues, runaway inflation and a growing labor shortage have hindered the industry's ability to increase output.** Oilprice.com. March 1, 2022.

"Oil prices could hit $240 per barrel this summer in the worst-case scenario if Western countries roll out sanctions on Russia's oil exports en masse. That's according to Rystad Energy's head of oil markets, Bjørnar Tonhaugen. Oilprice.com. March 10, 2022.

"**How much extra oil could OPEC+ pump to cool prices?** Saudi Arabia and the United Arab Emirates are among the few oil producers globally with spare capacity they could draw on quickly to increase output, help offset supply losses from Russia or elsewhere and ease prices, analysts say." Reuters. March 11, 2022.

"**[Russian President] Putin has weaponized energy,**" said [Senator Joe] Manchin. 'So, we have to have a better weapon, and that can take the form of U.S. energy development. We need energy independence….we have a responsibility as the leader of the free world.'" World Oil. March 11, 2022.

"Big Oil tax proposal - 50% 'windfall' tax rate to fund stimulus checks. **Legislators introduced a bill Thursday to tax the windfall profits of large oil companies at a 50% rate. Tax proceeds would be returned to consumers earning less than $75k/y through direct payments. The bill, as written, could create challenges and opportunities across the energy value chain."** SeekingAlpha. March. 11, 2022.

"Could Oil Prices Really Hit $200? Commodity prices, including oil and natural gas, are all over the place. Russia's aggression in Ukraine, and the subsequent retaliation from the West, have sent global supply chains into chaos. Oil prices could soar even higher if new sanctions are levied on Russia's energy industry. Metals prices are all over the place, with rapid rises to record levels followed by sharp falls and considerable anxiety — anxiety that has also applied to surging oil prices. Agricultural commodities have likewise seen an unprecedented surge in prices since the war in Ukraine broke out. **There is a risk of unrest in parts of the world where food prices have toppled governments in the past."** Oilprice.com. March 13, 2022.

In Texas, calls to boost U.S. oil production after Russian invasion run into hard realities. **Labor shortages, supply chain issues, hesitant financial backers and a frosty relationship with the Biden administration have limited how much Texas oil and gas companies are ramping up production.** The Texas Tribune. March 25, 2022.

Rationing Looms As Diesel Crisis Goes Global. Russian refiners cut processing rates of diesel fuel. Already tight diesel supply is getting even tighter. Vitol's chief executive Hardy: diesel supply shortage could trigger rationing in Europe. Oilprice.com March 27, 2022.

Ample Canadian crude pipeline, rail export capacity exists if output rises. Canadian pipeline bottleneck ended Q4 2021. More pipeline, rail and barge projects in progress. Canadian volumes ramping up, but not dramatically. S & P Global. March 24, 2022.

F. Other Difficulties

Batteries On A "Ridiculous" Cost Surge, Chinese EV Maker Says. Battery prices for electric vehicles are seeing a "ridiculous" increase on the back of soaring raw material prices, and carmakers will soon have to raise prices if they haven't done it yet, according to the top executive of Chinese electric vehicle manufacturer Li Auto Inc. OilPrice.com March 21, 2022.

Ukraine War Drives up Cost of Wind, Solar Power. 'Greenflation' problems are particularly acute in U.S., where tariffs targeting China helped increase project costs, led to delays before Russian attack. Russia's invasion of Ukraine is further driving up the price of renewable-energy projects, which were already facing supply-chain strains and raw-materials increases before the war. WSJ. March 27, 2022.

Surging Oil Prices Could Spark A Global Recession. Federal Bank of Dallas economists warn that a global economic downturn may be unavoidable if a large portion of Russian energy exports remain off the market throughout the year. Billionaire investor Carl Icahn also warned that there could be a recession amid the surging inflation. Oilprice.com. March 26, 2022,

G. The Political Winds

"'OPEC has had a long-standing policy of not changing production or supply on the basis of geopolitical events,'" explained Hasan Alhasan, a research fellow on Middle East policy for the International Institute for Strategic Studies in London. 'They only change in response to changes in market fundamentals.'" DW.com. April 3, 2022.

"**France is to introduce a rebate of 0.15 euros ($0.16) per litre of transport fuel to help drivers cope with soaring pump prices,** Prime Minister Jean Castex said in an interview with daily newspaper Le Parisien." Reuters. March 12, 2022.

"Sky-high oil prices resulting from a potential Russian oil import ban could force **governments to pour more cash into fossil fuel subsidies** to shield consumers from rising energy bills, rather than use the money to fight climate change." Reuters. March 8, 2022.

Germany's Scholz says **energy independence means higher costs**. Reuters. March 27, 2022.

H. Conclusions

Sadly, to varying degrees what's happened recently has happened before. As writer and philosopher George Santayana is purported to have said, **"Those who do not learn history are doomed to repeat it."**

Acknowledgements

I gratefully thank my publisher, Rob Godwin. He's been a pleasure to work with, and his assistance and patience are greatly appreciated. Of special note are Roger Bezdek and Bob Wending, who I've worked with off and on for decades on a variety of the subjects covered in this book.

Finally, I gratefully acknowledge the host of people who've contributed to my understanding of energy issues over the years. Special thanks go to Dr. Jim Schlesinger for his wisdom, guidance, and friendship. He is missed. Others of note include the following: Kjell Aleklett, Steve Andrews, Colin Campbell, Ralph Carabetta, Doug Chapin, Bob Epperly, Mikael Höök, Sadad al-Husseini, Ken Kern, Jean Laharrere, Jerry Kulsinski, Mike Ramage, Ramy Shanny, Matt Simmons, Chris Skrebowski, Randy Udall, Tom Whipple, and Jim Zucchetto. To all, thank you.

While I've consulted with various people on aspects of this book over the years, it should be clear that what's written here is my work, except as noted.

Index

A
Adaption 36
ARCO 45, 50, 79

C
Carbon dioxide 28
Carpooling 113
China 3, 30, 34, 38, 56, 60, 74, 95, 144, 146
Coal 21, 64, 65, 119, 120, 123, 124, 137
Conservation 63
Coupon(s) 105, 108, 109, 111

D
Department of Defense (DOD) 37, 67, 69, 132
Development 35, 43, 145
Drake, Edwin 19, 20, 44

E
EIA 55, 56, 60, 61, 137
Ethanol 98, 99, 120
Exploration 43, 44, 45, 63, 131, 134
ExxonMobil 52, 60, 79, 80, 131

F
Farnsworth, Philo T. 20, 21, 148
Forecasting 7, 75
Fusion 140, 141

G
Gasoline 105, 106, 108, 109, 110, 111
GDP 88, 108, 109
Geoengineering 35, 36, 72
Germany 35, 123, 147

H
Heavy Oil 65, 120, 122
Hydrogen 101, 102

I
IEA 58, 118, 132, 133
India 3, 34, 95
IOC 60, 62
IPCC 24, 25, 26, 27, 29, 31, 32, 33, 36, 38, 70, 75

M
Marconi 20
Methane 28, 94, 96
Mitigation, oil 5, 63, 64, 65, 67, 71, 103, 105, 107, 109, 111, 113, 115, 117, 119, 121, 122, 123, 124, 125, 126, 127, 129

N
Nuclear power 139

O
Oil shale 64
Oil shortages, 1973 & 1979 104
OPEC 54, 55, 57, 58, 61, 77, 78, 79, 81, 132, 135, 145, 147

P
Peak oil 61, 132, 133
Pharmaceutical 18
Plastic 17
Prudhoe Bay 49, 50, 51

R
Rationing 103, 104, 146
Rebate 111
Refining 51
Reserve 83
R/P 83

S
Schlesinger, James R. 5, 73, 147
Shale oil 56, 64

T
Tax 106, 146
Television 20

U
UK 82, 132

V
Venezuela 53, 54, 78, 122, 123, 144

W
Water vapor 25

Biography of Robert L. Hirsch, Ph.D.

Dr. Hirsch's professional experience is in research, development, and commercial applications of energy technologies. He has done research and managed technology programs in oil and natural gas exploration and production, petroleum refining, synthetic fuels, fusion, fission, renewables, defense technologies, chemical analysis, and basic research.

Previous positions:

- Senior Energy Program Advisor, SAIC (World oil production)
- Senior Energy Analyst, RAND (Various energy studies)
- Vice President of the Electric Power Research Institute. (All aspects).
- Vice President and Manager of Research and Technical Services for Atlantic Richfield Co. (Oil and gas exploration and production technology).
- Founder and CEO of APTI (Commercial & Defense Department technologies).
- Manager of Exxon's synthetic fuels research laboratory.
- Manager of Petroleum Exploratory Research at Exxon. (Refining R & D).
- Assistant Administrator of the U.S. Energy Research and Development Administration (ERDA) for renewables, fusion, geothermal and basic research.
- Director of fusion research at the U.S. Atomic Energy Commission and ERDA
- Project Leader on Inertial Electrostatic Fusion Research at ITT Farnsworth
- Research staff member, Atomic International, SNAP Reactor Program

He has served on and chaired a wide range of advisory committees for government and industry. He holds 16 patents and has authored over 50 publications. He is past Chairman of the Board on Energy and Environmental Systems of the National Research Council, the operating arm of the National Academies of Sciences and has served on a number of National Research Council study committees.

His recognitions include National Associate of the National Academies, first recipient of the University of Illinois NPRE Distinguished Alumni Award, University of Michigan Alumni Society Merit Award, Eminent Engineer of Tau Beta Pi engineering honorary, University of Illinois Alumni Honor Award for Distinguished Service in Engineering, American Nuclear Society Outstanding Achievement Award, Distinguished Fellow of the Center for Theoretical Studies, University of Miami, Fusion Power Associates Leadership Award, ERDA Special Achievement Award, AEC Distinguished Service Award, William A. Jump Memorial Foundation Meritorious Award, & AEC Special Fellowship in Nuclear Science and Engineering.